Mindfully Facing Climate Change
Endorsements:

Reactions to the concept of 'climate change' today are many and various, similar to the ways one can react to the diagnosis of a threatening disease—denial, anger, anxiety and numbness are common—but it doesn't need to be this way. In this humble and thoughtfully presented book the author skillfully draws upon such classical Buddhist teachings as 'The Discourse on Mindfulness', 'The Elephant's Footprint' and 'The Seven Suns', as well as modern scientific studies, to show how this instability in the world can be a source of liberating insight rather than of fear or rage. In addition, through citing numerous pragmatic teachings and meditation techniques, the book shows us how we can change our habits and live in ways that conduce to the well-being of the planet, of our human community and of us as individuals. Climate change is indeed a challenging diagnosis but it can cause us to wake up, let go and unselfishly help the world—to live more fully and skillfully—which is a blessing indeed for all concerned.

Ajahn Amaro

This short book by Bhikkhu Anālayo, one of the world's leading authorities on early Buddhism, is unique among works on Buddhism and environment for its extensive attention to early Buddhist texts. His skillful exegesis of these passages provides lessons for our perilous age. He shows how Buddhists have connected morality to both environmental degradation and recovery; how we can use Buddhist perspectives and practices to deal with denial, anger, resignation, and other emotions provoked by contemplating climate change; and not only how Buddhism could help with the mitigation of or adaptation to climate change, but how a practitioner could use the challenges we face to grow spiritually. A new story for the world requires that we change our minds, and Anālayo shows how Buddhist methods for cultivating mindfulness and practicing compassion can help us mitigate climate change and adapt to its progression.

Daniel Cozort

Rather than immediately focusing on climate change directly, Bhikkhu Anālayo first explores early Buddhism's perspective on the Earth and environment within the context of the four Noble Truths, based on selected early Pāli texts and their Chinese parallels. His thoughtful set of reflections then culminate in the chapter on the Eightfold Path, 'Walking the Path', in which he focusses on various ways in which the Dharma is helpful in both facing climate change without losing one's equanimity through such things as anger or resignation, and being mindful of ways in which one can reduce one's carbon footprint through one's everyday actions, and skillfully work to influence others to do likewise.

Peter Harvey

Bhikkhu Anālayo's book is an extraordinary achievement. Thoroughly familiar with early Buddhism both as a scholar and as a practitioner, he is at the same time fully aware of the imminent disastrous consequences of human-made climate change. Starting from the four noble truths as a scaffolding for his presentation and from mindfulness as a central Buddhist practice, he analyzes the present situation and points out how a Buddhist should react and contribute to prevent further deterioration. To an admirable degree, Bhikkhu Anālayo succeeds in suggesting creative answers while faithfully preserving the spirit and the thought of early Buddhism.

Lambert Schmithausen

Mindfully Facing
Climate Change

Bhikkhu Anālayo

Published by
Barre Center for Buddhist Studies
Barre, Massachusetts
USA

(www.buddhistinquiry.org)

ISBN 9781706719885

The cover shows the stūpa at the Barre Center for Buddhist Studies; hornets have built a nest right above the head of the Buddha statue.

Cover picture taken by Benjamin Engel.

Cover design and drawings by Dido Dolmen.
(www.didodolmen.it)

As an act of Dhammadāna, Bhikkhu Anālayo has waived royalty payments for this book.

Contents

Acknowledgement

I am indebted to Chris Burke, Bhikkhunī Dhammadinnā, Ann Dillon, William Edelglass, Linda Grace, Peter Harvey, Karin Meyers, Yuka Nakamura, Mike Running, and Lambert Schmithausen for commenting on a draft version of this book and to the staff, board members, and supporters of the Barre Center for Buddhist Studies for providing me with the facilities needed for my practice and writing.

Foreword

As a Dharma teacher who is very concerned about climate, I have been aware for some time that practitioners are often resistant to hearing Dharma talks on this subject. I have been told at times that the topic is "too political" and "off point"—they have come to hear the teachings to be comforted in order to address their own personal suffering and hardly want to be asked to take on even more. They understandably seek to escape from the intensity and bombardment of sensational news filled with acrimony, divisiveness, and fear that one must work hard to avoid. But I wonder about the contradiction of engaging in practice while avoiding the truth.

While wanting to offer words that can calm the mind and heart, I also take to heart the Buddha's example that in order to truly be free we need to directly acknowledge *dukkha* so we can skillfully transform it into compassion and wisdom. I wrestle with finding the balance between words that can soothe and support in facing the climate crisis and those that challenge us to wake up to the fact that we are like children playing with our toys without realizing that the house is on fire. So with deep appreciation I commend Bhikkhu Anālayo for addressing in *Mindfully Facing Climate Change* this crisis we are all facing on this planet.

Sitting with Bhikkhu Anālayo one can't help but be impressed and inspired by his knowledge of not only the Pāli discourses but their Chinese and Tibetan parallels as well. Even more, the inspiration that comes from witnessing his deep practice is invigorating. So it was with enthusiasm and profound respect that I sat with him at Spirit Rock on a weeklong retreat exploring the intricacies of the *Ānāpānasati-sutta*.

However, I was completely unprepared for the teachings he gave on the last morning of that retreat. Sitting rock steady in the Dharma hall as usual, he began by saying, "This morning I am going to talk about climate change. Some people think that this is not an appropriate topic for the Dharma Hall. I want you to know that I think it is an important thing we should be talking about in our practice." A deep hush came over the room.

He then proceeded to give a powerful explanation about why this topic is so crucial and what our meditation practice and Buddhist teachings have to offer as we navigate this crisis facing all of humanity. I had no idea that this had been a topic of deep concern for this Buddhist scholar and meditation teacher. It gave me hope that someone so steeped in the teachings and such a respected voice of the Theravāda tradition was applying the Dharma in this way.

At this time, with so many of us wondering how we can wisely approach the crisis of climate change, it is a real gift to see how Bhikkhu Anālayo applies the teachings as a way to hold this dire situation. He shows us how Dharma understanding and perspective can help not only with our personal practice but also can contribute profound wisdom for humanity as we all grapple with the situation.

In these pages he explores in his own unique way how various canonical teachings such as the four noble truths, the four elements, the *brahmavihāra*s (with a special emphasis on compassion), mindfulness, the eightfold path, and death contemplation can guide us in addressing the climate issue. What is particularly refreshing is Bhikkhu Anālayo's way of creatively applying the teachings and canonical discourse references to the current situation. Furthermore, along with theoretical applications, he offers accompanying meditative practices that deepen an embodied understanding of these concepts and help us develop the equanimity and inner resilience needed to skillfully meet this enormous challenge facing all of us.

It could be easy to believe that a monk who has such clarity and depth of practice does not struggle or deal with strong emotions when considering this issue. But this is not the case. When he asked if I would write some words as a foreword to this book, I said I would gladly consider it, but I had one request: that I could let readers know that he too struggles with holding the pain of the climate situation.

In a conversation with him I recorded for a Spirit Rock day-long on the climate crisis, he related that he had chosen to make his concerns about climate the focus and theme of a recent five-week self-retreat. Each morning he would contemplate the truth of the facts about the impact of global warming and consider how his practice could hold it all. What a challenge to confront and absorb the pain of the world and possible demise of human life on the planet in that way!

Bhikkhu Anālayo is not one to talk about his personal experience. But when pressed, he confessed that he too was dealing with the type of reactions that can get activated with regard to the situation. I believe it is important for Dharma practitioners to see that even someone so practiced in equanimity and balance has to work hard in order to be able to process the almost incomprehensible threat to life on this planet.

This crisis is something that will demand from us, as a species, greater wisdom, compassion, courage and emotional stability than humanity as a whole has ever needed. As Bhikkhu Anālayo makes abundantly clear in this book, the Buddha's teachings and their application to our changing climate are perhaps some of the most significant tools we can offer society to help us rise to this challenge. May reading this book support you in your own life and inspire you to be an agent of positive change for us all in this endeavor.

James Baraz
Berkeley, California
October 17, 2019

Introduction

This booklet serves as a companion to a freely available online course,[1] intended to offer a practical approach to the challenges of climate change that is grounded in the teachings of early Buddhism.[2]

Human-caused climate change and environmental destruction are modern problems. These were unknown at the time of the Buddha; hence it can hardly be expected that the early discourses offer precise advice on how to handle these. Nevertheless, several early discourses provide helpful perspectives and can be relied on in facing the current challenge.

As a scaffolding for my presentation of relevant textual passages, I rely on the four noble truths. The first chapter takes up the first truth, based on a discourse that expounds this topic in relation to the four elements, the first of which is earth. The second chapter uses the lens of the second truth to examine the causes that have led to the current predicament, in particular the three unwholesome root defilements (greed, hatred, and delusion) and their relation to climate change denial and other unwholesome reactions.

The third chapter relates to the third truth with a focus on the divine abodes (*brahmavihāra*) as wholesome mental states

---

[1] See https://www.buddhistinquiry.org/analayo.

[2] By way of reflecting the importance accorded to early discourse material, I depart from a standard way of textual formatting by instead setting off all canonical passages as block quotations (in a slightly larger font size), no matter how short they may be. Conversely, I do not adopt block quotations for material from secondary sources, even when these are of considerable length, but quote them inline.

of temporary liberation that have considerable ethical potential. Out of these four divine abodes, compassion is particularly relevant to the environmental situation. In the fourth chapter, I apply the eightfold path to the problem of climate change, based on attempting to show that facing this problem can offer a substantial contribution to the overarching goal of this eightfold path: progress to awakening.

In line with the basic thrust of the eightfold path, a central concern in my presentation in these four chapters is to try to develop a practice-related approach to the climate crisis. A key element here is a form of meditation that attempts to put the main themes of the four chapters of this booklet into practice.

In an appendix, I survey this meditation practice, which combines mindfulness directed to the internal and external earth element with an examination of the condition of the mind, a cultivation of compassion, and contemplation of impermanence. The form of meditation that results from combining these four practices is meant to help build up resilience and wisdom for facing the dire consequences of climate change.

Assessing the teachings of early Buddhism requires comparing parallel versions of a particular discourse, as at times during oral transmission portions of text may have been lost or else added. Whereas in the online course I rely on the Pāli version of the relevant passages, in this booklet I provide translations of the respective Chinese parallels, which often are as yet untranslated.[3] As a rule of thumb, material found similarly in parallel versions provides a good foundation for reconstructing early Buddhist thought.

---

[3] In order to enable readers to make their own comparison of the parallels without needing to consult the Pāli original, in the footnotes I refer to the standard translations of the relevant Pāli texts.

The overall approach I present here has no pretense of being the only right one. When in the course of my discussion I explain why certain ideas do not align with my understanding of the early Buddhist perspective, this is not meant to encourage their wholesale rejection. My point is only that these are not part of the approach chosen here, which after all is just one out of many ways of trying to respond to the current predicament. The challenge of climate change is of such magnitude that it calls for a range of different perspectives, which can and should complement each other under the common aim of minimizing harm and ensuring, to the best of our ability, the sustainability of life on this planet.

Relating to the Earth

The crux of the problem of climate change is how we, as human beings, relate to the environment. As a way of approaching this relationship from an early Buddhist perspective, in this chapter I try to cover three main topics. These three involve offering an introduction to some key doctrines of Buddhism, surveying meditative approaches to the earth, and attempting to establish an ecological ethics in line with the teachings reflected in the early discourses.

I begin by briefly sketching some key doctrinal teachings of early Buddhism that are of relevance, in one way or another, for the remainder of this and the following chapters. These are the four noble truths, the significance of *dukkha*, the purpose of analyzing subjective experience into five aggregates, the relationship of the human body to the earth, and the teaching on dependent arising.

Out of these doctrinal teachings, the dependency of the human body on the earth is of particular relevance to my overall exploration. I study this dependency based on the Discourse on the Elephant's Footprint, which relates the internal earth element, found inside one's own body, to the external earth element outside.

The Discourse on the Elephant's Footprint also serves as a lead-in to the second topic, featuring as the first of three discourses, studied in this chapter, that offer meditation-related perspectives on the earth element. The other two discourses connected to this topic depict how to contemplate the earth element as a mindfulness exercise and how to take the earth as an example for cultivating an attitude of equanimity.

The last of the three topics to be taken up in this chapter explores how to ground ecological concerns in early Buddhist thought. This is far from straightforward and has been an ongoing topic of scholarly discussion.[1] The challenge here is that the overall concern of early Buddhism is progress to liberation. In contrast, nature as such is not invested with a value in and of itself (Schmithausen 1997a).

It follows that environmental concerns, as long as these are developed from within the context of early Buddhist thought, need to be established in a way that avoids relying on the idea that ecosystems have a value on their own and for this reason should be protected. In the course of trying to build a proper foundation for an environmental ethics concordant with early Buddhism, I need to survey critically selected positions taken by some Green Buddhists.

As part of my overall attempt to do justice to the teachings of early Buddhism, I have adopted what is perhaps its most central doctrinal teaching as a scaffolding for this book as a whole: the four noble truths. Each of the ensuing chapters begins with a quote, outlining one of these four truths. The quotes are from one of the Chinese parallels to the Discourse on Turning the Wheel of Dharma (*Dhammacakkappavattana-sutta*), traditionally regarded as the first teaching delivered by the Buddha after his awakening. In keeping with this approach, I now turn to the first truth.

The Four Noble Truths

The first of the four truths takes the following form:[2]

---

[1] For surveys of different positions see Harris 1994 and Swearer 2006.

[2] EĀ 24.5 at T II 619a10, parallel to SN 56.11 at SN V 421,19 (translated by Bodhi 2000: 1844); for comparative studies of SN 56.11 see Anālayo 2012b and 2013a.

What is reckoned to be the truth of *dukkha*?[3] That is, birth is *dukkha*, old age is *dukkha*, disease is *dukkha*, death is *dukkha*, grief and vexation are *dukkha*, worry, sadness, and afflictions that cannot be measured [are *dukkha*]; association with what is disliked is *dukkha*, dissociation from what is liked is *dukkha*, not getting what is wished for is also *dukkha*; stated in brief, the five aggregates of clinging are *dukkha*. This is reckoned to be the truth of *dukkha*.

The early Buddhist teaching of four truths appears to be modeled on an ancient Indian scheme of medical diagnosis (Anālayo 2011c). This scheme covers a diagnosis of the disease, an identification of what is responsible for the disease, a determination of the potential for recovering health, and a prescription of the required cure. Based on this precedent, the teaching on these four noble truths can be understood to involve recognition of the following:

1) *diagnosis*: the stressful repercussions of whatever difficulty one is facing,
2) *etiology*: how one is contributing to that distress through craving and attachment,
3) *prognosis*: the potential for reducing distress by cultivating a different attitude,
4) *treatment plan*: the path of practice to be undertaken to achieve a change of attitude and a reduction of distress.

The diagnosis of the first truth, in the passage translated above, lists various instances of "*dukkha*". Before exploring this di-

---

[3] A noteworthy difference, compared to SN 56.11, is that EĀ 24.5 does not apply the qualification "noble" to each of the four truths; see in more detail Anālayo 2006.

agnosis further, a few words are required about the import of this term.

Dukkha

The term *dukkha* has been regularly translated as "suffering". This translation, however, fails to capture adequately the different dimensions of this term in its early Buddhist usage.

One of these dimensions is the experience of what is painful, where *dukkha* stands for one of the three feeling tones (*vedanā*), the other two feeling tones being pleasant and neutral. In the case of experiencing pain, this does not invariably have to result in suffering. Through training in mindfulness, it becomes possible to face the challenge of pain with a balanced mind (Anālayo 2016d). Hence *dukkha* as one of the three feeling tones can refer to "pain" or at least what is "unpleasant", but this does not invariably have to result in "suffering".

Another dimension of the same term concerns all conditioned phenomena, which can without exception be qualified as *dukkha*. This usage thereby covers all three feeling tones: pleasant, painful, and neutral.

Pleasant experiences could hardly be reckoned "suffering". Of course, pleasant experiences eventually change and their disappearance can be quite frustrating. But painful experiences also change, and in that case the change will probably be experienced as welcome. Therefore, the fact of change can also not unequivocally be considered as productive of suffering.

This goes to show that "suffering" is not a quality shared by all conditioned phenomena. Instead, it is a reaction of an untrained mind. For this reason, applying the term "suffering" as a qualification to all conditioned phenomena fails to make sense. As a qualification of conditioned phenomena in general, a translation as "unsatisfactory" would be more to the point.

In order to avoid misleading connotations, in the remainder of my presentation I will continue to employ the Pāli term *dukkha*, in the hope that with the above brief exploration I have given sufficient background for appreciating its significance.

The Five Aggregates

The translation given above from a Chinese version of the Discourse on Turning the Wheel of Dharma agrees with its Pāli parallel in reckoning the five aggregates of clinging as *dukkha*. The scheme of five aggregates involves an analysis of subjective experience into the following aspects:

1) bodily form,
2) feeling tone,
3) perception,
4) volitional formations,
5) consciousness.

These five can be understood to point to the material, affective, cognitive, conative, and sentient dimensions of subjective experience. The purpose of this analytical scheme is to highlight five modalities of clinging to a sense of self. The body comes to be clung to as "where" I am, feeling tones as "how" I am, perceptions as "what" I cognize, formations as "why" I act in a certain way, and consciousness as "whereby" I experience.

According to the summary statement of the first truth, all of these five dimensions of subjective experience are *dukkha*. Such assessment reflects the early Buddhist soteriological orientation mentioned at the outset of this chapter. In fact, the formulation of the first truth starts off with birth as a manifestation of *dukkha*. Human birth offers a valuable opportunity insofar as it provides suitable conditions for progress on the path to liberation, but it is not seen as valuable in and of itself. The final aim of the path to liberation is precisely to transcend any type of birth.

With the final goal of the early Buddhist path attained, arahants are considered to be completely free from clinging to any of the five aggregates and therewith from future birth. Even encountering old age, disease, and death no longer leads to "grief and vexation" or to "worry, sadness, and afflictions" in them.

At the same time, from the arahant's viewpoint all conditioned phenomena are still *dukkha* in the sense of being incapable of providing true and lasting satisfaction. From an early Buddhist perspective, *dukkha* is an inherent quality of nature,[4] however broadly this latter term is defined.

The Elephant's Footprint

For a further assessment of the first truth, I turn to a discourse that has as its speaker a chief disciple of the Buddha by the

---

[4] *Pace* Holder 2007: 121, who in the context of a criticism of Schmithausen 1997a reasons that "to say that the things of the world are suffering in and of themselves (i.e., independently or intrinsically) merely on account of their being impermanent is an unwarrantedly pessimistic interpretation of the early texts" as "natural phenomena are characterized as *dukkha only in the context of an experiential relationship that relates certain natural phenomena to an unenlightened being having sentience*." According to the canonical passage John Holder quotes in support of this position, SN 36.11 at SN IV 216,23 (translated by Bodhi 2000: 1271), all that is felt is included in *dukkha*. The discourse continues by explaining that this statement refers to the impermanence of all formations. This is thus the canonical position, rather than being an "unwarrantedly pessimistic interpretation". Whatever is felt remains impermanent and therefore unsatisfactory even for one who has reached full awakening.

name of Sāriputta. This is the Discourse on the Elephant's Footprint, which begins by highlighting the comprehensive role of the four truths. The presentation in this discourse is of particular relevance to my overall concerns, as it establishes a relationship between the first of the five aggregates, the human body, and the earth.

The highlight on the four truths in a Chinese version of this discourse takes the following form:[5]

> Friends, it is just like animal footprints, of which the foremost is an elephant's footprint. Why is that? It is because the elephant's footprint is the largest.
>
> Friends, it is like this with countless wholesome states; all these states are completely contained in the four noble truths, they fit into the four noble truths, and the four noble truths are reckoned foremost among all states.
>
> What are the four? They are reckoned to be the noble truth of *dukkha*, of the arising of *dukkha*, of the cessation of *dukkha*, and the noble truth of the path to the cessation of *dukkha*.
>
> Friends, what is the noble truth of *dukkha*? That is, birth is *dukkha*, old age is *dukkha*, disease is *dukkha*, death is *dukkha*, association with what is disliked is *dukkha*, separation from what is loved is *dukkha*,[6] not getting what is wanted is *dukkha*; in short, the five aggregates of clinging are *dukkha*.

---

[5] MĀ 30 at T I 464b23, parallel to MN 28 at MN I 184,26 (translated by Ñāṇamoli 1995/2005: 278); for a comparative study of MN 28 see Anālayo 2011b: 193–198.

[6] MN 28 does not explicitly mention disease and the *dukkha* of association and separation, although these are mentioned in other Pāli discourse formulations of the first truth, such as, e.g., SN 56.11 at SN V 421,20 (translated by Bodhi 2000: 1844).

The image of the elephant's footprint occurs also in other discourses. In several such instances it serves to highlight the importance of diligence (*appamāda*).[7] Alternatively, the same imagery can illustrate the importance of either the perception of impermanence or else of the faculty of wisdom.[8] This last usage would fit the present context well, as it must be the wisdom potential of the four noble truths that earned it a position of eminence comparable to the footprint of an elephant.

According to Ñāṇaponika (1966/1981: 2), the present passage conveys that "the Four Noble Truths comprise ... all that is beneficial, i.e. all that is truly worth knowing and following after." Cousins (1996: 146) adds that when "all skilful dhammas are included in the four noble truths, we should ... interpret skilful dhammas here as referring to meditational states. Indeed, this is made clear later in the *sutta* by the references to equipoise connected with the skillful."

A meditative perspective, related to the cultivation of insight, does indeed emerge in the course of the ensuing exposition in the Discourse on the Elephant's Footprint. Before coming to that, however, the discourse first proceeds from the four noble truths to the earth element.

The Four Elements

The exposition in the Discourse on the Elephant's Footprint lists the five aggregates individually in order to explain the statement that "the five aggregates of clinging are *dukkha*" and then takes up just the first aggregate of bodily form. After ex-

---

[7] SN 3.17 at SN I 86,31 (translated by Bodhi 2000: 179) and its parallel MĀ 141 at T I 647c6; SN 45.140 at SN V 43,13 (translated by Bodhi 2000: 1551); AN 6.53 at AN III 364,22 (translated by Bodhi 2012: 926); AN 10.15 at AN V 21,18 (translated by Bodhi 2012: 1354); SĀ 882 at T II 222a5; and SĀ² 66 at T II 396b27.

[8] SĀ 270 at T II 70c15; SN 48.54 at SN V 231,2 (translated by Bodhi 2000: 1697).

plaining that bodily form is made up of the four elements of earth, water, fire, and wind, the analysis proceeds by taking up just the first element of earth, representative of the quality of solidity. In this way, the exposition gradually leads from the teaching of the four noble truths to a close inspection of the nature of the earth element. This takes the following form:[9]

> What is the earth element? Friends, the earth element is reckoned to be of two types: there is the internal earth element and there is the external earth element.
>
> Friends, what is the internal earth element? It is reckoned to be what is internal, being found inside the body, what is contained internally and is solid, established in the nature of solidity, being internally clung to. And what is that?
>
> It is reckoned to be head hair, body hair, nails, teeth, rough and smooth epidermis, skin, flesh, sinews, bones, heart, kidneys, liver, lungs, spleen, intestines, stomach, feces, and whatever else similarly exists in this way inside the body, is contained internally [and is solid], established in the nature of solidity, being internally clung to. Friends, this is reckoned to be the internal earth element.

The identification of the solid parts of the body as earth element can be interpreted by recourse to another passage, found in a Pāli discourse and its Chinese parallel. The passage in question describes a skilled meditator who is able to see the wood of a tree as a manifestation of the earth element, or of the water element, or of the fire element, or else of the

---

[9] MĀ 30 at T I 464c5, parallel to MN 28 at MN I 185,14; this part of the discourse has also been preserved as a quotation in Tibetan, Up 3044 at D 4094 *ju* 140b6 and P 5595 *tu* 161b5.

wind element, as each of these four elements is found in the wood.[10]

Applying the same perspective to the passage translated above, the other three elements must also be found in the bodily parts listed. On reflection, this is indeed the case. Each of these bodily parts has some degree of cohesion or liquidity, corresponding to the water element, each has some warmth, reflecting the fire element, and within each part of the body some motion takes place, indicative of the presence of the wind element.

It follows that these bodily parts are listed as manifestations of the earth element in order to highlight that this element is predominant; the above passage would not imply that they are only earth and nothing else. These bodily parts are predominantly solid and thereby suitably illustrate the nature of the internal earth element.

In the Discourse on the Elephant's Footprint, this is the first step to be implemented, before turning to the external earth element in the form of what is outside of the body and similarly solid. Amaro (2013: 22) explains that "the first place to explore our relationship to the *earth element* is on the individual level. This body—with its bones and blood, its hair, teeth, and organs—is the bit of the Earth that we can know most directly, most acutely, and most intimately."

The Earth

The Discourse on the Elephant's Footprint turns to the external earth element by describing how it can be affected by flooding:[11]

Friends, there are times of deluge, and at such times the

---

[10] AN 6.41 at AN III 340,28 (translated by Bodhi 2012: 904) and SĀ 494 at T II 128c23.

[11] MĀ 30 at T I 464c11, parallel to MN 28 at MN I 185,27.

external earth element disappears [under water]. Friends, this external earth element, which is very great, very clean, and very non-repulsive, is [still] of an impermanent nature, of a nature to cease, of a nature to decline, and of a nature to change; how much more so is this transient body, which is clung to with craving.

Yet, one reckoned an unlearned foolish worldling has this thought: "This is me; this is mine; I belong to it." A learned noble disciple, [however], does not have this thought: "This is me; this is mine; I belong to it." How could one have such a thought?

The overall thrust of the passage is to drive home the impermanent nature of one's own body and hence the meaninglessness of clinging to it as an embodiment of "me" or "mine". In order to make this point, the passage establishes the similarity in nature between the body and the external earth. Given that even the external earth is impermanent, the body must also be impermanent.

The depiction of the disappearance of the earth reflects an ancient Indian cosmological belief that the world goes through cyclical periods of destruction, as a result of which human beings will disappear from the earth. Such destruction could be due to water, as in the present case, or else because of fire or wind.[12] I will take up a description of such destruction based on fire in the last chapter of my study (see below p. 120).

After having related the internal to the external earth element, the Discourse on the Elephant's Footprint continues by showing how insight cultivated in this way fortifies one when

---

[12] See the *Abhidharmakośabhāṣya*, Pradhan 1967: 189,1 (translated by Pruden 1988: 490) or the *Visuddhimagga*, Vism 414,13 (translated by Ñāṇamoli 1991: 410).

experiencing abuse. The same insight can inspire the cultivation of equanimity to such an extent that one will even be able to face a physical attack with inner balance. The discourse applies the same procedure to the other three elements of water, fire, and wind.

The overall exposition in the Discourse on the Elephant's Footprint combines two themes of relevance to meditative culture of the mind, namely insight into the absence of a self and the growth of equanimity. I will return to the part of the discourse related to equanimity in a later chapter (see below p. 91).

The above description of the disappearance of the external earth is at the same time about the disappearance of human life from the earth. Once the whole earth is inundated, the foundation for the survival of human beings would indeed have disappeared. The Discourse on the Elephant's Footprint thereby implicitly drives home the fact that the existence of the body depends on the existence of the earth.

The relationship established in this way offers a starting point for developing an ecological concern that aligns with early Buddhist thought. The presentation in the Discourse on the Elephant's Footprint implies that the external and internal earth element share the same impermanent nature. It also implies that the impermanent nature of the earth affects the body, which must perish if the earth comes to destruction. Clearly, the continuity of the body depends on the continuity of the earth.

Such dependency is quite evident in the need to nourish the body. Besides food as a manifestation of the earth element, the body also requires regular intake of the water element in the form of beverages, maintenance of a certain temperature range, corresponding to the fire element, and a continuous supply of oxygen through the process of breathing, a manifestation of the wind element.

The body's need for nourishment forms part of a canonical description in the following form:[13]

> The material body, which [consists] of the four elements and the six sense-faculties, which has been given birth to and been raised by father and mother, grows up from milk and meals.

The Pāli counterpart does not refer to the six sense-faculties or to the parents.[14] Instead of milk and meals, it just speaks of feeding on solid food. Alongside such minor differences, both descriptions agree that the body consists of the four elements and needs to be fed.

The conditionality that emerges in this way would be sufficient reason for a concern with protecting living conditions on earth. DellaSalla (2020: 100) sums up the situation as follows: "humans need Nature to thrive ... but the reverse is certainly not true. Nature does not need humans to persist or thrive ... Whatever we do to Nature, we ultimately do to ourselves, so protecting the natural world and abiding by what it needs to thrive will be in the long-term interest of humanity."

Protecting the natural world indeed safeguards the basis for sustaining the human body, which in turn is needed for a human being to be able to cultivate mindfulness, and such

---

[13] DĀ 28 at T I 110c14, parallel to DN 9 at DN I 186,1 (translated by Walshe 1987: 163). A Sanskrit fragment parallel, Melzer 2006: 260 (418v1), only mentions that the body is solid and made up of the four elements.

[14] Other instances of descriptions in Pāli discourses of the nature of the human body do explicitly recognize the role played by the parents, although they tend to mention the mother first. An example is MN 74 at MN I 500,1 (translated by Ñāṇamoli 1995/2005: 605); for a comparative study of MN 74 see Anālayo 2011b: 399–406.

cultivation is required for progress on the path to awakening.[15]

A Pāli discourse without parallels lists scarcity of food as one of several potential dangers that could occur in future times, highlighting that such scarcity will lead to migration of people and result in a situation where it becomes rather difficult to practice the Buddha's teaching.[16] The main purpose of this description is to instill a sense of urgency (saṃvega) so that one practices with wholehearted dedication as long as food is still readily available. At the same time, the passage also points to a practitioner's dependency on suitable living conditions.

The body's dependency on nutriment also comes up in an exposition concerning the four establishments of mindfulness (satipaṭṭhāna). The discourse in question depicts the arising and ceasing of the object taken by the first establishment of mindfulness, the body, in this way:[17]

> The arising of nutriment is the arising of the body; the cessation of nutriment is the vanishing of the body.

This indication in a way invites mindful inspection of the fact that the body to be contemplated in formal meditation on the four establishments of mindfulness depends on nutriment.

---

[15] This reasoning should hopefully meet the criterion presented by Harris 1994: 46 in the following form: "The minimum qualification for an authentic Buddhist ethics is that it is able to construe causation in such a way that goal-oriented activity makes sense."

[16] AN 5.78 at AN III 104,9 (translated by Bodhi 2012: 711). AN 5.54 at AN III 66,6 (translated by Bodhi 2012: 681) confirms that a time of food scarcity is not suitable for making an effort at the practice.

[17] SĀ 609 at T II 171a29, parallel to SN 47.42 at SN V 184,17 (translated by Bodhi 2000: 1660), a Tibetan parallel, Up 6031 at D 4094 nyu 15a5 or P 5595 thu 48b5 (translated by Dhammadinnā 2018a: 23), and an Uighur fragment, Zhang 2002: 109.

Without nutriment, the body ceases, and therewith the object of the first establishment of mindfulness disappears.

Contemplation of the Elements

One of the exercises for cultivating the first of the four establishments of mindfulness is directly relevant to a meditative approach to the earth element. This approach can serve to reveal the meaninglessness of clinging to the body as an embodiment of "me" or "mine", also thematized in the Discourse on the Elephant's Footprint. The relevant passage takes the following form in a Chinese parallel to the Discourse on the Establishments of Mindfulness (*Satipaṭṭhāna-sutta*):[18]

> A monastic contemplates this body by distinguishing the elements in this body as being the four elements. This is just like a capable cow butcher or the apprentice of a cow butcher who divides a cow [into pieces by cutting through] its tendons. While dividing it he contemplates and sees for himself that "these are the feet," "this is the heart," "these are the tendons," and "this is the head".

The purpose of the element contemplation is to lead to a dissection of one's sense of selfhood by driving home the fact that what is experienced as "my" body is, after all, just a combination of the four elements.

The shift of perspective to be achieved in this way is similar to a butcher who, after having slaughtered a cow and cut it

---

[18] EĀ 12.1 at T II 568a25, parallel to MN 10 at MN I 57,35 (translated by Ñāṇamoli 1995/2005: 148), see also DN 22 at DN II 294,14 (translated by Walshe 1987: 338), and parallel to MĀ 98 at T I 583b18, which expands from four to six elements; for a comparative study of MN 10 see Anālayo 2011b: 73–97 and 2013b, the latter of which comes with full translations of EĀ 12.1 and MĀ 98.

up, no longer thinks that "this is a cow," as his perception of the cow as a compact unit has ended. Instead, he thinks in terms of pieces of meat, which he is putting out for sale.[19]

This helps to draw out the implications of this particular mindfulness exercise of contemplating the earth element, which serves as a way of putting into practice the realization of the empty nature of the human body, in line with the chief teaching of the Discourse on the Elephant's Footprint.

A Mind Like the Earth

In addition to the Discourse on the Elephant's Footprint and the Discourse on the Establishments of Mindfulness, a third discourse relevant to developing a meditative approach to the earth, to be examined in this chapter, takes up each of the four elements as an inspiration for cultivating equanimity. This is a topic also thematized in the Discourse on the Elephant's Footprint. The relevant instruction for the case of the earth element proceeds as follows:[20]

> You should maintain your mind like the earth. Just as this earth receives what is pure and also receives what is impure, it receives excrement and urine, and all that is dirty and disgusting, yet the earth does not give rise to a discriminatory mental attitude; it does not say: "This is

---

[19] Although this purpose is not expressed explicitly in MN 10 in a manner comparable to EĀ 12.1, the same idea is reflected in the corresponding commentary, Ps I 272,1.

[20] EĀ 43.5 at T II 760a5. A comparable instruction can be found in MN 62 at MN I 423,18 (translated by Ñāṇamoli 1995/2005: 529). Although the parallel to this discourse, EĀ 17.1 at T II 582a13, does not have the corresponding instruction, it seems quite probable that this is due to a shifting of the relevant textual portion from EĀ 17.1 to EĀ 43.5; see in more detail Anālayo 2014/2015: 76f.

attractive, this is repulsive." Now your practice should
also be like this.

This passage encourages taking inspiration from the imper-
turbability of the earth in order to develop a mind that similarly remains unperturbed.

The meditative perspectives that emerge from the three discourses taken up in this chapter point in complementary ways
to the themes of emptiness and equanimity. Here the Discourse
on the Elephant's Footprint provides the foundational perspective on the similarity in nature of the internal and the external
earth element, thereby undermining the tendency to consider
the human body as separate and special. The empty nature of
the human body emerges in particular in its utter dependency
on what is external to it.

A practical way to explore this empty nature of the human
body can be found in mindful contemplation of the elements.
The present passage offers additional support for what has already emerged from these discourses by presenting the earth
as a model for cultivating mental balance, in order to remain
unperturbed in the face of the vicissitudes of life.

The Earth and Sentience

In addition to providing this practice-related perspective, the
passage on keeping the mind like the earth also offers an indication relevant to environmental ethics. Readers less interested
in a detailed discussion of this topic might prefer to skip the
remainder of this chapter and just read the summary on page
48 below.

The reference to what is dirty and loathsome being dumped
on the earth links the passage translated above to ecology, even
though this was of course not the intention of the original dis-

course. Of further significance is the encouragement to remain free from reactivity when facing agreeable and disagreeable experiences, as it relies on the notion that the earth does not react. The passage thereby gives the impression that the earth and the other elements were not seen as endowed with sentience.

Of relevance to this topic is also a rule of monastic conduct, which forbids digging the earth. Seeing monastics engaged in digging the earth had provoked censure by people who did perceive the earth as sentient,[21] a view apparently common in the ancient Indian setting. The ruling presumably reflects the need to avoid offending public opinion.[22] In fact, a monastic text extant in Chinese, the Mahāsāṅghika *Vinaya*, reports the Buddha stating explicitly that the earth is not endowed with the faculty of sentience, before promulgating the rule against digging it oneself or getting it dug by others.[23]

The famous scene when, on the eve of his awakening, the

---

[21] The *Suttavibhaṅga*, Vin IV 32,25, reports the censure as follows: "How can recluses, [followers] of the son of the Sakyan, dig the earth or have it dug? The recluses, [followers] of the son of the Sakyan, are harming sentience with a single faculty" (i.e., the sense faculty of touch).

[22] Schmithausen 1991: 57 concludes his survey of relevant material regarding "the belief that earth is by itself a kind of living, sentient being (or aggregate of such beings)" by leaving it open whether "this belief was, as the *Suttavibhaṅga* puts it, only the belief of 'people' which Buddhist monks and nuns were enjoined to take into account in their behaviour, or whether it was, in the earliest period, still shared by the Buddhist monks and nuns themselves, or had at least not yet been abandoned by them on a conscious or theoretical level."

[23] T 1425 at T XXII 384c16: "it is devoid of the faculty of sentience," in evident reference to the earlier mentioned "earth"; a regulation already noted by Schmithausen 2009: 42 note 72.

Buddha called the earth to witness, is an element of later hagi-
ography and not attested in the early discourses.[24] Even this
later stage, however, does not involve the idea of anthropomorphizing the earth as a mother, an image sometimes in-
voked to inspire environmental protection.[25]

Considering the earth as a mother, in the sense of viewing
the production of food and living conditions as something
done intentionally to benefit sentient beings, does not seem to
be successful in providing a coherent grounding for environmental concerns. If the earth is endowed with intentionality,
then volcanic eruptions, earthquakes, and tsunamis, with all
their catastrophic repercussions, would also have to be attributed to intentional acts of the earth.

Once such intentionality is granted, in order to stimulate a
debt of gratitude to mother earth for her abundant gifts, it can
hardly be avoided that such destructions are equally viewed as
the earth's intentional deeds. Intentional destruction on a massive scale like this, however, does not fit the image of a mother
particularly well.

In order to motivate environmental concerns, a different
approach would seem to be required, at least as long as such
concerns are to be grounded in early Buddhist thought.

Humans and Animals

The Discourse on the Elephant's Footprint does provide a starting point for such a different approach in order to situate ecological ethics within the context of early Buddhist thought.
However, such relating an environmental concern to the dependency of humans on an intact ecosystem, in order to be

---

[24] Jā I 74,26 (translated by Jayawickrama 1990: 98).
[25] On this idea see also Dhammadinnā 2018b: 300.

able to practice the path to awakening, remains within an anthropocentric perspective.

Although this would suffice as a rationale for environmental action when living conditions on earth are endangered, it will probably not satisfy those who consider the privileging of humans over other species to be the chief culprit for the current predicament. In order to address this issue, I now survey some relevant observations.

Regarding the topic of the hierarchical positioning of humans in relation to animals, Light (2002: 428) explains that "many early environmental ethicists were adamant that, if environmental ethics was going to be a distinctive field of ethics, it necessarily had to involve a rejection of anthropocentrism in ethics." In this context, "the notion of what anthropocentrism meant, and in consequence what overcoming anthropocentrism entailed, often relied on very narrow, straw-man definitions of anthropocentrism. Anthropocentrism was equated with forms of valuation that easily, or even necessarily, led to nature's destruction" (2002: 429).

"As a consequence, those agents of change who will effect efforts at environmental protection—namely, humans—have oddly been left out of discussions about the moral value of nature. As a result, environmental ethics has been less able to contribute to ... the resolution of environmental problems" (2002: 427).

The problem here is simply that such an "environmental ethics appears more concerned with overcoming human interests than redirecting them toward environmental concerns. As a result, a nonanthropocentric form of ethics has limited appeal" to the general audience, "and not to appeal to such an audience arguably means that we are not having an effect either on the formation of better environmental policies or on the project of engendering public support for them" (2002: 436).[26]

Light (2002: 444) concludes that "a more fully responsible environmental ethics must abandon the wholesale rejection of anthropocentric reasons for protecting the environment."

The need to avoid a wholesale rejection of anthropocentrism holds not only for environmental ethics in general but also for the more specific case of Buddhist approaches to ecology. Sponberg (1997: 361) explains that "we have identified in our own cultural history an unquestionable tendency toward attitudes of exploitation and domination of nature, and we have rightly associated those attitudes with cultural institutions of hierarchy and privilege.

"The unwitting and often quite unconscious mistake we make, however, comes when we assume that all forms of hierarchy are the same. We assume that any and every manifestation of hierarchy leads inevitably to the dead end of dominion and exploitation ... and, as Western Buddhists, we reassure ourselves that any apparently hierarchical element in our cherished Buddhism must be a mistake, perhaps the later corruption of some monastic elitists."

Although "Buddhism advocated, in its early form at least, a ... decentralized institutional structure, this should not be misconstrued in the light of our current Western concerns to mean that the spiritual ideal in Buddhism was seen as nonhierarchical and egalitarian" (1997: 352).

"Some Green Buddhists, uncomfortable with any notion of hierarchy ... are moving, intentionally or not, toward a kind of

---

[26] In fact, even mobilizing political action to counter climate change based on invoking human rights is not necessarily straightforward and poses challenges; see Jodoin et al. 2019.

unidimensional Buddhism" (1997: 371). "Obsessed with the need to dump out the dirty bath-water of Western hierarchies of oppression, some Green Buddhists fail to notice that they are also discarding the 'baby'" (1997: 372).

That some form of hierarchy need not invariably be exploitative emerges also in a comment by Harvey (2000/2005: 150f), who explains that "humans are 'superior' primarily in terms of their capacity for moral action and spiritual development. [However], the natural expression of such 'superiority' is not an exploitative attitude, but one of kindness to lesser beings."

The moral responsibility that results from a hierarchically superior position has been highlighted by the Dalai Lama (2009: 24), who reasons that "morally, as beings of higher intelligence, we must care for this world. Its other inhabitants— members of the animal and plant kingdoms—do not have the means to save or protect it. It is our responsibility to undo the serious environmental degradation caused by thoughtless and inappropriate human behavior."

In short, accepting the idea of some hierarchical superiority of human beings does not mean that compassion and care based on a moral sense of responsibility are being diminished.

The Food Chain

Turning from secondary sources to the early discourses, the reasoning underpinning a consideration of animals as inferior to humans finds expression in the following manner:[27]

---

[27] MĀ 199 at T I 761b23, parallel to MN 129 at MN III 169,22 (translated by Ñāṇamoli 1995/2005: 1021) and T 86 at T I 909a13, both of which take up the lower destinies of rebirth together and give less detailed descriptions. A later text, the *Saddharmasmṛtyupasthāna-sūtra*, even incorporates such assessment of the nature of the an-

> Among animals there is no practice of altruism and justice, no practice of the principles of moral conduct, no practice of what is sublime and wholesome. Those animals eat up one another. The strong eat the weak; the big eat the small.

According to this passage, it is precisely the lack of moral agency that, from an early Buddhist perspective, makes animals appear inferior to humans. This lack is particularly evident in the food chain. In this way, often enough "to be an animal in a Buddhist cosmos is to live a miserable and pathetic existence" (Ohnuma 2017: 5).

As already noted by Schmithausen (2000: 72), an "affirmation of the natural world as it is implies affirmation of its basic structures, among which the *food chain* is the most scandalous one because it involves an awful amount of killing and pain."[28] From this viewpoint, there would be little room left for considering animals as in themselves sacred.[29] Taking into account the repercussions of the food chain could hardly instill veneration or be considered in some way sanctified, unless one were to redefine the notion of sacredness completely, as a result of

---

imal world in its version of the fourth establishment of mindfulness, contemplation of dharmas; see Stuart 2019: 5.

[28] The same would also be of relevance to what Harris 1995a: 201 considers an instance of "terminological revisionism", which "repositions the central Buddhist term *saṅgha* by moving it away from its traditionally monastic domain so that it may act as a designation for the totality of all beings."

[29] Harris 1997b: 380 argues that "environmentalism—particularly of the ecospiritual type, a form that has a sizable impact on contemporary ecoBuddhism—represents a reappropriation of prescientific modes of thinking with its ... insistence on ... an almost pantheist power of nature."

which it would no longer be able to function as a motivation for environmental activism. In other words, to endorse fully the killing done by animals would substantially weaken the ethical basis for censuring the destruction caused by humans.

In the case of plants, the early discourses do reflect the ancient Indian belief that these can be inhabited by spirits (Schmithausen 2009: 77–83). Hence cutting down a tree, for example, might incur the wrath of the spirit that dwells in it. However, in modern times it seems less probable that such beliefs still have sufficient influence to promote forest preservation.[30] Besides, even in the ancient Indian setting such beliefs do not involve attributing sacredness to the trees in themselves.

Of interest here is a discourse which describes how someone might fell a great tree, cut it into pieces, dry these, burn them up in a fire, and then throw the ashes into a river.[31] This description serves to convey positive connotations. The point made in this way is that, just as this tree has no scope to arise again, in the same way *dukkha* has no scope to arise again in the case of one who has reached full awakening.

A similar attitude is also evident in the simile employed to convey the gist of mindful contemplation of the earth, taken up

---

[30] In the case of Thailand, for example, the loss of such beliefs in modern times seems to be a major reason for increasing deforestation; see Seeger 2014: 52–54.

[31] SN 12.55 at SN II 88,5 (translated by Bodhi 2000: 591), parallel to Sanskrit fragments, Tripāṭhī 1962: 88f (§2.7&8), and SĀ 284 at T II 79c14; see also SHT X 3865 Vw, Wille 2008: 200, which has preserved a reference to a swift current into which the ashes are thrown (not preserved in the fragments edited by Tripāṭhī 1962). The relevance of SN 12.55 as a contrastive example to a tendency among some Buddhist environmentalists to attribute inherent value to trees, because the Buddha attained awakening seated under one of them, has already been noted by Schmithausen 1997a: 70 note 178.

earlier, which describes a butcher cutting up a cow for sale (see above p. 32).

Both cases of course do not actively encourage the slaughter of an animal or the destruction of a plant. Nevertheless, the image of their destruction is employed to convey a positive nuance. Had plants and animals been considered sacred, such a depiction would hardly have been chosen to illustrate a commendable form of meditative practice or attainment.

Dependent Arising

The above survey of problems related to construing an ecological ethics is certainly not meant to dismiss forms of environmental activism that are based on a biocentric approach. The point is only that, in order to stay within the framework of early Buddhist thought, the pragmatic approach I attempt to present here needs to rely on an anthropocentric concern with the environment rather than attributing an intrinsic value to animals or nature.[32] This is simply because, as already noted by Swearer (2001: 231f), "Indic Buddhism was certainly not biocentric." At the same time, as already mentioned in the introduction, what I present here is just one of many possible approaches to achieve the common goal of reducing the dire effects of climate change.

The anthropocentric concern adopted here is based on the

---

[32] Eckel 1997: 343 comments on the proposal of an "intrinsic value of animals, plants, rivers, mountains, and ecosystems" that "the word 'intrinsic' presents a barrier. It seems to suggest precisely the substantial, permanent identity that the ideas of no-self and interdependent co-origination are meant to undermine" (see also Rockefeller 1997: 320). Yet, the problem is not the proposal that some quality is intrinsic. In fact, the teaching on not self proposes an intrinsic quality shared by all phenomena.

understanding that for human beings to walk the path to liberation requires basic living conditions, which in turn are dependent on the natural environment.

The dependency of the human body on the earth as the mo-
tivator for environmental activism could then be considered an
instance of specific conditionality, the principle underlying the
early Buddhist teaching on dependent arising (*paṭicca samup-
pāda*).

The Discourse on the Elephant's Footprint, taken up earlier
in this chapter, in fact explicitly connects its presentation to
dependent arising. Its exposition culminates in the following
statement:[33]

> Friends, the Blessed One has also spoken in this way: "If
> one sees dependent arising, one in turn sees the Dharma; if
> one sees the Dharma, one in turn sees dependent arising."
>
> Why is that? Friends, the Blessed One taught that the
> five aggregates of clinging arise from conditions: the form
> aggregate of clinging, the feeling tone, the perception, the
> formations, and the consciousness aggregate of clinging.

In this way, the discourse moves from the individual elements
back to the level of the five aggregates of clinging as a summary statement of the first truth. The whole intervening exposition was a detailed breakdown of the first of these aggregates,
and the present passage clarifies that this served to exemplify
its dependently arisen nature.

The early Buddhist doctrine of dependent arising concerns
specific conditions and their cessation; it does not establish a
general interconnectedness of all things, comparable to sys-

---

[33] MĀ 30 at T I 467a9, with its parallel in MN 28 at MN I 190,37
(translated by Ñāṇamoli 1995/2005: 283).

tems theory (Schmithausen 1997a: 13). The notion of a general interconnectedness of all things, as a basis for inspiring action to preserve a harmonious balance in nature, differs from the purpose of the early Buddhist teaching on dependent arising, whose aim is rather to step out of conditionality. Be it the dependently arisen five aggregates of clinging or the dependency of the body on the external earth element, the ultimate aim of such teachings is complete transcendence rather than a positive evaluation of these conditions.

Sucitto (2019: 249) comments, on the promotion of the interconnection of all things to motivate ecological concerns, that in this "shift from the perspective of the early suttas, this state of interconnection is to be encouraged, rather than constituting a description of what one needs to be released from. This was a distinct change of meaning from that of the [early] Buddhist tradition."

Besides, it is also not clear how, on being invested with a positive evaluation, the notion that everything is interrelated could provide a coherent basis for an ecological concern. Harris (1995b: 177) points out that "the intention here is to show that since all things are inter-related we should act in a spirit of reverence towards them all. However, the category of 'all things' includes insecticides, totalitarian regimes and nuclear weapons and the argument therefore possesses some rather obvious problems. In short, it suffers from a certain vacuity from the moral perspective."

In the same vein, Ives (2013: 563) comments on the promotion of non-duality as the complement to interconnectedness: "haunting Buddhist celebrations of wholes is the naturalistic fallacy, as seen when celebrations of the whole undermine the ability to make the sort of distinctions that are necessary and unavoidable in environmental ethics, such as the distinction between a negative 'is' (such as a toxic river) and a positive 'ought' (the clean river that can result from clean-up efforts)."

As noted similarly by Markus et al. (2018: 20), "a preference for 'good' things is not yet given through the premise of non-dualism alone, rather further principles are needed to establish the ethical foundation."

In sum, from an early Buddhist perspective it seems difficult to ground ecological concerns coherently by anthropomorphizing the earth as a mother whose benevolent care needs to be recompensed, by viewing animals and plants as sacred, or by relying on the notion that everything is interrelated.

The Beauty of Nature

The above considerations regarding the early Buddhist perspective on the earth do not imply that there cannot be an appreciation of the beauty of nature. Such appreciation occurs indeed regularly in the discourses, which de Silva (1990: 15) explains as expressing a "contemplative attitude by which we discern in nature our own vision of peace and tranquility ... emerging from this contemplative attitude, there is an aesthetic dimension that reinforces our move toward conservation. There are many references in the Buddhist texts to instances where persons of great spiritual heights appreciate scenic beauty ... [and are] able to look at the mirror of nature without attachment and with equanimity."

One example is a discourse which takes the beauty of a flowering forest as its starting point. The discourse begins with some fellow monastics approaching Sāriputta for a Dharma discussion. He starts off the discussion by drawing attention to the delightful place in which they had met, which forms a recurrent theme throughout their exchange.

According to the Chinese version of the discourse in question, Sāriputta formulates this theme in the following manner:[34]

---

[34] MĀ 184 at T I 727a18, parallel to MN 32 at MN I 212,31 (translated by Ñāṇamoli 1995/2005: 307), EĀ 37.3 at T II 710c15, T 154.16 at T III 81a15, and a reference to *ramaṇīyam* in a Sanskrit fragment parallel, Or.15009/422, Hirabayashi 2015: 291; for a comparative study of MN 32 see Anālayo 2011b: 209–216.

This *sal* forest at Gosiṅga is highly delightful,[35] the night is brightly moonlit, and the *sal* trees are spreading a delicate fragrance, as if from divine flowers.

The beauty of the moonlit forest, with the flowers in full bloom, leads on to a discussion on what particular spiritual quality could match this beauty. Each of the monastics present on this occasion describes one particularly praiseworthy quality, such as dwelling in seclusion, mastery of the mind, exemplary moral conduct, etc. In this way, the aesthetic dimension of the beauty of nature acquires its true significance from a soteriological perspective.

Another example would be a verse in a collection of early Buddhist poetry, which proceeds as follows:[36]

Delightful are forest wilds,
Here ordinary folk will not delight;
The passionless find delight here,
Those who are not in quest of sensual pleasures.

The charm of forest wilds relates to freedom from sensual desire. This holds to such an extent that those who have reached

---

[35] The *sal* tree, *shorea robusta*, is "a majestic tree growing up to 45 metres in height and having a girth of 3.6 metres, with ovate oblong leaves and pale yellow flowers" (Dhammika 2015/2018: 179).

[36] The *Udānavarga* 29.17, Bernhard 1965: 375, with parallels in Dhp 99 (translated by Norman 1997/2004: 15; see also Th 992, translated by Norman 1969: 92), the Patna *Dharmapada* 155, Cone 1989: 143, T 210 at T IV 564b14 (translated by Dhammajoti 1995: 146), T 211 at T IV 588c23 (translated by Willemen 1999: 93), T 212 at T IV 749c29, and T 213 a T IV 793b6 (translated by Willemen 1978: 132). On *araṇya* occurring in a range of texts to convey the sense of wilderness, in contrast to areas cultivated by humans, see Visigalli 2019.

this lofty goal are fully able to delight in them, unlike those caught up in the quest for the pleasures of sensuality.

As noted by Sandell (1987: 47), the verse shows "how advancement of mind leads to a greater appreciation of nature." Sucitto (2019: 182) explains that "the core meaning is that when the awakened mind reviews the manifest world, it exhibits a deep appreciation of 'uncivilized' nature."

The condition of the mind emerges as the crucial element in relation to the appreciation of natural beauty and the attitude toward the environment in general. In the next chapter, I will examine this topic in detail.

Summary

From an early Buddhist viewpoint, animals and nature are not invested with value in and of themselves. The earth is not conceived as a mother, and the teaching on dependent arising is not an affirmation of the interrelationship of all things.

An approach to environmental concerns that wishes to stay in line with the teachings of early Buddhism can therefore best be developed based on an anthropocentric perspective, in the sense that the human body requires appropriate living conditions on earth in order to be able to serve as a vehicle for progress to awakening. Such an anthropocentric paradigm comes with a recognition of the moral responsibility of humans toward other sentient beings.

From a meditative perspective, central themes that emerge in relation to the earth element are emptiness and equanimity. These can be cultivated through a mindful contemplation that proceeds from the internal earth element, found inside the body, to its external counterpart.

An Ethics of the Mind

What is the truth of the arising of *dukkha*?[1] That is, it is grasping conjoined with craving that leads to acting carelessly with a mind that keeps being lustfully attached. This is reckoned to be the truth of the arising of *dukkha*.

In this chapter I explore the causes underlying the *dukkha* of environmental destruction and climate change.

Of central importance to the present chapter is the Discourse on the World Ruler, which provides an ethical perspective on environmental decline. In early Buddhist thought, such ethical perspective has the mind as a central reference point. Hence, following up the implications of the Discourse on the World Ruler leads me to the topic of mindful contemplation of the mind and to exploring common responses to the challenge of climate change from the perspective of their relation to mental defilements.

The Arising of *dukkha*

The statement on the second truth, translated above, highlights the chief role that craving plays in the early Buddhist analysis of the conditions responsible for *dukkha*.[2] This concords with a central thrust in Buddhist thought, which is a keen interest in what is taking place in the mind. Throughout, the emphasis is less on what is out there, but much rather on what takes place within, which is in turn evaluated from an ethical perspective.

---

[1] EĀ 24.5 at T II 619a14.
[2] On this relationship see in more detail Anālayo 2019a.

The ethical perspective afforded by the second truth can conveniently be related to environmental concerns: "the realization is that maximization of affluence and output derived from economic activity ... with high levels of social and natural environmental impact ... will fail to close the desire–want ↔ satisfaction gap. Furthermore ... discontent will be exacerbated by a value system where goals and expectations are incorrectly predicated on the idea that well-being is a function of increasing consumption" (Daniels 2010a: 958).

In the end, as pointed out by Harvey (2013: 336), "a consumer economy actually depends on people not being satisfied with the goods they buy, or at least not satisfied for long."

Mindful Self-reliance

For exploring the causes leading to environmental decline, the Discourse on the World Ruler provides a helpful perspective with its depiction of a gradual deterioration of living conditions in ways that are similar to current predictions of the dire repercussions of climate change. Throughout the discourse, such deterioration is directly linked to a decline in morality.

The description itself is part of what appears to be a parable,[3] whose purpose is to illustrate the importance of building self-reliance through the cultivation of mindfulness. A Chinese version of the Discourse on the World Ruler sets out this main theme, which the ensuing description serves to flesh out, in the following manner:[4]

---

[3] On the nature of this description as a parable or myth see, e.g., Rhys Davids and Rhys Davids 1921: 53, Gombrich 1988: 83, and Collins 1998: 481.

[4] DĀ 6 at T I 39a24, parallel to DN 26 at DN III 58,7 (translated by Walshe 1987: 395) and MĀ 70 at T I 520b19; a full translation of DĀ 6 can be found in Anālayo 2014a.

You should be a light unto yourselves, with the Dharma as your light, without any other light; you should be a refuge unto yourselves, with the Dharma as your refuge, without any other refuge.

How will monastics be a light unto themselves, with the Dharma as their light, without any other light; be a refuge unto themselves, with the Dharma as their refuge, without any other refuge?[5]

Here monastics contemplate the body as a body internally, being diligent without laxity, with undistracted mindfulness, removing greed and sadness in the world. They contemplate the body as a body externally ... they contemplate the body as a body internally and externally, being diligent without laxity, with undistracted mindfulness,[6] removing greed and sadness in the world. They contemplate feeling tones ... mental states ... dharmas *also in this way*.[7]

This is how monastics are a light unto themselves, with the Dharma as their light, without any other light;

---

[5] The Chinese text does not explicitly indicate plural forms, so that it would have been more natural to translate it in the singular as "a monastic". My adoption of plural forms is due to the wish to avoid gendered terminology.

[6] The translation "undistracted mindfulness" is based on adopting a textual variant reading, in the sense of an alternative formulation preserved in some editions of the text, which in the presence case concords with the formulation found earlier in the discourse; the original instead additionally refers to "consciousness".

[7] DN 26 at DN III 58,11 describes *satipaṭṭhāna* meditation without bringing in a distinction between internal and external modes of practice. MĀ 70 at T I 520b23 proceeds directly from the injunction to be a lamp unto oneself, without referring to *satipaṭṭhāna*, which in this version comes only at the end of the discourse.

they are a refuge unto themselves, with the Dharma as their refuge, without any other refuge.

The reference to being "a light" to oneself in the above passage reflects one of two possible understandings of the underlying Indic term, which has its Pāli counterpart in *attadīpa*. Alternatively, this term could also be understood to convey the sense of being "an island" to oneself.[8]

The imagery of being a light or island to oneself recurs in other contexts, the majority of which relate the practice of mindfulness to being confronted with something distressing.[9]

One such instance concerns the passing away of Sāriputta, the speaker of the discourse taken up in the last chapter. Learning of his death had triggered considerable sorrow in Ānanda, the monastic who served for many years as the personal attendant of the Buddha. According to a Chinese version of the relevant discourse, Ānanda described his own condition as follows:[10]

Now my whole body is [as if it were] falling apart, the four directions [are as if they had] changed their order, the teachings I learned are [as if they were] blocked off.

The Pāli version similarly reports him stating that his body felt

---

[8] Norman 1990/1993: 87 explains that the term "could mean either 'a lamp for oneself' or 'an island, i.e., refuge, for oneself' ... either *ātma-dīpa* or *ātma-dvīpa*"; see also the discussion in, e.g., Bapat 1957, Brough 1962/2001: 210, Nakamura 2000: 95, and Wright 2000.

[9] An instance where this is not the case nevertheless comes with an encouragement to examine the causes for sorrow and pain; see SN 22.43 at SN III 42,8 (translated by Bodhi 2000: 882) and its parallel SĀ 36 at T I 8a22 (translated by Anālayo 2014c: 9).

[10] SĀ 638 at T II 176c8, parallel to SN 47.13 at SN V 162,15 (translated by Bodhi 2000: 1643).

as if he had been drugged, he had become disoriented, and the teachings were no longer clear to him. To help him overcome this condition, the Buddha gave Ānanda a teaching culminating in the recommendation to be a light or island to oneself by cultivating the four establishments of mindfulness. Clearly, here mindfulness practice is the tool to emerge from grief and sorrow.

Another instance takes its occasion from the fact that, after Sāriputta's death, his close friend Mahāmoggallāna, another disciple of considerable importance in the monastic community, had also passed away. In a teaching delivered on this occasion, the Buddha acknowledged that the assembly of his disciples now appeared empty, due to the demise of these two chief disciples. Nevertheless, his recommendation was to avoid giving rise to sorrow on this account:[11]

> How could it be that what is of a nature to be born, of a nature to arise, of a nature to be constructed, of a nature to be conditioned, of a nature to change, will not be obliterated? The wish to make it not become destroyed is for something that is impossible.

This clarification then leads over to the recommendation to become a light or island to oneself through mindfulness practice.

A further occurrence relates to the Buddha's own potential passing away. He had been so seriously ill that Ānanda was worried that his teacher was about to pass away. A Chinese version of this episode reports his sentiments as follows:[12]

---

[11] SĀ 639 at T II 177a28, parallel to SN 47.14 at SN V 164,18 (translated by Bodhi 2000: 1645).

[12] DĀ 2 at 15a23, parallel to DN 16 at DN II 99,22 (translated by Walshe 1987: 244) or SN 47.9 at SN V 153,10 (translated by Bodhi 2000: 1636); for a comparative study of this part of the *Mahāparinirvāṇa-*

When the Blessed One was ill, my mind was in fear and tied up with worry. I felt lost in bewilderment, no longer recognizing the directions.

Here, once again, the Buddha's reply leads up to the recommendation to become self-reliant, a light or island to oneself, by cultivating mindfulness. This advice culminates in the declaration that, by dint of such cultivation, one becomes a true disciple of the Buddha:[13]

After my final Nirvana, those who are able to cultivate this teaching are truly my disciples and foremost in the training.

The Pāli parallels express a similar meaning by stating that those who become self-reliant through mindfulness will be foremost among those who are keen on training.

Based on the passages surveyed above, it can be anticipated that the key theme of the Discourse on the World Ruler will similarly be about the potential to face grief and sorrow with mindfulness, thereby learning to become self-reliant.

Although this is indeed the case, the Discourse on the World Ruler does not take its occasion from the death of a close one. Instead, it depicts a gradual moral and environmental decline that culminates in truly catastrophic conditions on the earth. Nevertheless, the same advice holds in this case: become self-reliant through mindfulness. It follows that the same advice also holds in relation to the main topic of this book, in that mindfulness is indeed the central tool for facing climate change.

---

sūtra that takes into account the other extant parallels see Waldschmidt 1944: 88–94.

[13] DĀ 2 at 15b14, parallel to DN 16 at DN II 101,1 or SN 47.9 at SN V 154,15.

Bad Governance

The Discourse on the World Ruler relates the onset of a gradual deterioration of the environment to bad governance. After the peaceful reign of several kings who maintained ancient customs, a king assumes power who instead follows his own whims:[14]

> This one king governed the country on his own; he did not continue the ancient law. His government was unstable, everyone was complaining, the country was declining, and the people were withering away.
>
> Then one brahmin minister approached the king and said: "Great king, you should know that the country is now declining and the people are withering away. Things are not turning out as usual. Now the king has many good friends in the country who are wise and erudite, knowledgeable in things ancient and modern. They are equipped with knowledge of how earlier kings governed rightly by the Dharma.[15] Why not command them to gather and ask what they know, so that they will give their personal replies?"
>
> Then the king summoned his many ministers and asked them about the way earlier kings had governed. Then the wise ministers provided answers on these matters. The king heard what they said and implemented the old way of governing and protecting the world by means of the Dharma. However,[16] he was unable to aid solitary elderly people as well as to provide for the lowly and destitute.

---

[14] DĀ 6 at T I 40b15, parallel to DN 26 at DN III 64,27 and MĀ 70 at T I 521b25.

[15] The translation "rightly" is based on adopting a variant reading.

[16] The translation "however" is based on adopting a variant reading.

Another Chinese version of the present discourse offers further information on what had made this particular king govern according to his personal whims:[17]

> Yet, he was stained by sensual pleasures, attached to sensual pleasures, insatiably greedy for sensual pleasures, in bondage to sensual pleasures, affected by sensual pleasures, and dominated by sensual pleasures. He did not see their disadvantage and did not know a way out of them. So he ruled the country according to his own ideas. Because he ruled the country according to his own ideas, the country consequently declined and no longer prospered.

Although this detailed explanation is peculiar to the above discourse, all versions report that the king failed to look after the poor. Hence, the suggestion in the above passage that sensuality was responsible for his neglect and bad governance fits the context well. As the king of the country, he would have had ample opportunities for sensual gratification. The resultant indulgence could indeed have fostered a neglect of the destitute.

A similar pattern is evident in the current environmental crisis, due to a pervasive concern with national interests among leaders in affluent countries.[18] This often results in a failure to

---

[17] MĀ 70 at T I 521b25.

[18] Of relevance to the impact of national concerns is also the observation by Deese 2019: 15 that "strategic competition among sovereign nation states has greatly impeded efforts to understand and address environmental challenges. In particular, the rise of militaristic nationalism has caused extensive environmental destruction and has frequently corrupted the practice of science by tying it to the secretive culture of the national security state." As a result, "the enduring competition between rival nation states is one of the reasons that the collective behavior of our species is less intelligent

take properly into account the situation in less affluent countries, where living conditions keep worsening due to the effects of climate change resulting directly from excessive indulgence by the wealthier.

The indication provided in the other Chinese version would also be in line with a general principle proposed by Sivaraksa (2009: 63), according to which in those times "when good governance erodes, one or more of the three poisons of greed, hatred, and ignorance is always present."

Ethical Decline

Not providing for the needy had its consequences, which the Discourse on the World Ruler describes as follows:[19]

> Then the people of the country in turn became quite impoverished. Consequently, they took from one another by force, and theft increased greatly. It being investigated, they seized one of them, took him to the king, and said: "This man is a thief. May the king deal with him."
>
> The king asked him: "Is it true that you are a thief?"
> He replied: "It is true. I am poor and hungry, unable to

---

than it could be" in addressing climate change (2019: 17). In the same vein, Lueddeke (2019: 84) observes that "water, energy, food security and climate increasingly require a search for pragmatic decisions to ensure the global survival of all species. Unfortunately, finding consensus around these survival issues is continually sidetracked and threatened by ongoing dangerous, human-centric, ideological, territorial and often self-centred battles at the highest levels—despite the fact that history has repeatedly demonstrated that there can be no winners in the long run and that genuine collaboration is the only way forward."

[19] DĀ 6 at T I 40b23, parallel to DN 26 at DN III 65,17 and MĀ 70 at T I 522a28.

maintain myself. Therefore, I have become a thief." Then the king supplied him with goods from his treasury and said: "With these goods support your parents and care for your relatives. From now on, do not become a thief again!"

Other people in turn heard that the king was giving wealth to those who engage in theft. Thereupon they further engaged in stealing the property of others.

The problem of theft originated from the king's failure to provide for the poor, and hence from his lack of compassion.[20] When faced with the results, the king tried to act in a compassionate manner. However, this was misguided compassion; it lacked wisdom, as the king failed to realize that it involved rewarding immoral behavior.

When thievery continued to occur again and again, the king realized the consequences of his misguided compassion and decided to punish the thieves:[21]

They again seized one of them, took him to the king, and said: "This man is a thief. May the king deal with him."

The king asked again: "Is it true that you are a thief?" He replied: "It is true. I am poor and hungry, unable to maintain myself. Therefore, I have become a thief."

Then the king thought: "At first, seeing that they were poor, I gave the thieves wealth so that they would stop. But other people have heard of it and in turn imitated each other even more, and robbery increases daily. This will not do. Let me now rather have that man pilloried. I

---

[20] Harris 1997a: 9 highlights that in this discourse "lack of compassion for the poor leads to the disintegration of society."

[21] DĀ 6 at T I 40c7, parallel to DN 26 at DN III 66,34 and MĀ 70 at T I 522b14.

will command that he [be paraded through] the streets and alleys and then taken out of the city to be executed in the wilds, as a warning to other people."

Then the king ordered his attendants: "Have him bound, beat a drum to announce the command,[22] and [parade] him through the streets and alleys. This done, take him out of the city and execute him in the wilds."

The people in the country all came to know that someone who had become a thief had been taken and bound by the king, who commanded that he [be paraded] through the streets and alleys and executed in the wilds. Then the people said to one another in turn: "If we are labelled as thieves, we will be like that, not different from him."

Then the people in the country, to protect themselves, consequently made themselves weapons to fight with, swords and bows with arrows. They repeatedly killed and injured each other when attacking to plunder.

From the time this king came [to the throne], poverty started. There being poverty, robbery started. There being robbery, fighting with weapons started. There being fighting with weapons, there was killing and harming. There being killing and harming, [people's] complexions became haggard and their lifespan shorter.

In agreement with its parallels, the above passage depicts an increasing decline in moral standards, where poverty leads via theft to mutual killing. Gombrich (1988: 84) reasons that "this text states that stealing and violence originate in poverty and that poverty is the king's responsibility; punishment becomes necessary only because of the king's earlier failure to prevent poverty. This humane theory, which ascribes the origin of crime

---

[22] The translation "beat" is based on adopting a variant reading.

to economic conditions … is not typical of Indian thinking on such matters … this idea is so bold and original that it is probably the Buddha's."

The other Chinese version again provides a detail that helps to appreciate the narrative flow. According to its report, the previously described thief had been caught by the owner.[23] This indication fits the flow of the narration well, as when the king decided to punish thievery, people took up weapons to kill those they robbed. This would be a logical consequence of thieves earlier being arrested by the owners, because by killing the owner(s) one could avoid being arrested and then punished by the king.[24]

The idea of a reduction of lifespan is a recurrent motif throughout the entire story, expressed in ways that reflect the nature of the tale as a parable. According to the overall progression, the initial life span is either eighty thousand or forty thousand years, and this gradually diminishes until reaching an all-time low of ten years.[25]

Of relevance for evaluating such details is the largely symbolic function of numbers in the early Buddhist discourses.[26] Vansina (1985: 171) explains that, in an oral setting in general, numbers "are both abstract and repetitive so that they fare badly in all [oral] traditions and are stereotyped to numbers meaning 'perfect', 'many', 'few'." In addition, in ancient times numbers possessed a significance of their own quite apart from their function as devices for mathematical calculation. In relation to another discourse, Syrkin (1983: 156) speaks of "an archaic and

[23] MĀ 70 at T I 522b1.
[24] In fact, MĀ 70 at T I 522b27 reports that people, on taking up arms, think that "if … we catch the owners of those goods, we will cut off their head."
[25] See the survey of these time periods in Anālayo 2014a: 17.
[26] For some examples see Anālayo 2011b: 471 note 158.

universal tendency to describe the world with the help of definite number complexes ... manifoldly reflected in the Pali canon."

In view of such symbolic function, perhaps the actual fig-
ures could be accorded less importance in order to discern the
main principle behind the narrative. A decline in average life
expectancy would be natural once burglary and killing set in.
Without intending to pretend that this must have been the original intent of the description, such an interpretation would help
to convey the basic message of the parable in a way more easily appreciated in modern times.

Ecological Decline

The moral decline also has ecological repercussions, which the
Discourse on the World Ruler depicts in this way:[27]

> At that time one no longer hears in the world the names
> of ghee, rock honey, dark rock honey, or of any sweet
> delicacies. Rice seeds and rice seedlings turn into grass
> and weeds. Silk, silken cloth, brocade, cotton, white wool,
> what now in the world is called a garment, are at that
> time not seen at all. Fabrics woven from coarse hair will
> be the best kind of clothing.
>
> At that time many thorny bushes grow on this earth,
> and there are many mosquitoes, gadflies, flies, fleas,
> snakes, vipers, wasps, centipedes, and poisonous worms.
> Gold, silver, lapis lazuli, pearls, and what are called gems
> completely disappear into the earth. On the surface of the
> earth there appear only clay stones, sand, and gravel.[28]

---

[27] DĀ 6 at T I 41a13, parallel to DN 26 at DN III 71,17 and MĀ 70 at
T I 523a13.

[28] The translation "only" is based on adopting a variant reading.

At that time human beings never ever hear any more the names of the ten wholesome [actions]. The world will be just full of the ten unwholesome [actions].[29] When [even] the names of the good qualities are no longer present, how could those people get to cultivate wholesome conduct?

At that time human beings are capable of being extremely bad and there is no filiality toward parents, no respect for teachers and elders, no loyalty, and no righteousness. Those who are rebellious and without principles are esteemed.[30] It is just as nowadays those are esteemed who are [instead] capable of cultivating wholesome conduct, of filiality toward parents, of respecting teachers and elders, of being loyal, trustworthy, and righteous, of following principles and cultivating compassion.[31]

At that time [human] beings recurrently engage in the ten bad [courses of action] and often fall into bad ways. On seeing one another, [human] beings constantly wish to kill one another.[32] They are just like hunters on seeing a herd of deer.

---

[29] On the ten unwholesome courses of action (*kammapatha*) see below p. 65.

[30] The translation is based on adopting two variant readings.

[31] The translation "compassion" is based on adopting a variant reading. The corresponding passage in DN 26 at DN III 72,2 describes esteeming those who respect parents, recluses, brahmins, and the elders of the clan; MĀ 70 at T I 523a19 refers to esteeming those who practice the ten wholesome courses of action.

[32] According to DN 26 at DN III 72,14 and MĀ 70 at T I 523a25, mutual hatred will even arise between close relatives (like mother and son, etc.). DN 26 at DN III 72,8 stands alone in showing the breakdown of family relationships as also affecting sexual conduct, in that people will copulate with each other without respect for one's mother, aunt, or teacher's wife.

Then on this earth there are many ravines, deep gorges with rushing rivers. The earth is a wasteland and people are scarce. People go about in fear.[33] At that time fighting and plundering will manifest, grass and sticks taken in the hand will all become [like] halberds and spears. For seven days they will turn to mutual harming.[34]

---

[33] The translation "people" is based on adopting a variant reading.

[34] DN 26 at DN III 73,4 explains that people will get a perception of each other as deer, *migasaññaṃ paṭilabhissanti*. This takes up the motif of the deer simile found in all versions to illustrate the hatred human beings have toward each other.

The above version stands alone in giving such a detailed description of the environmental decline. Although its parallels agree in depicting an absolute low point in living conditions and morality, they do not describe it in such depth. Thus the parallelism to the anticipated repercussions of current climate change (see Wallace-Wells 2019) emerges with such detail only in this particular version.

Other discourses help in understanding how in the ancient setting the repercussions of ethical decline were believed to affect the environment. One relevant passage occurs in a Pāli discourse with a parallel extant in Chinese. Similar to the Discourse on the World Ruler, this discourse also sets out from a condition of bad governance, whereupon wrong conduct gradually penetrates from the highest to the lowest ranks of society. Once everyone behaves in unrighteous ways, this affects the heavenly constellations and leads to storms:[35]

> Then storms manifest. Storms having manifested, the celestials become upset. The celestials having become upset, at that time wind and rain become untimely. At that time grains and seeds in the earth no longer grow.

The two parallels agree that, as a final result, people become short-lived. The reference to celestials (*devas*) in both versions reflects the ancient Indian belief that such beings are responsible for rain.

According to a Pāli discourse without parallels, one of the reasons for a lack of rainfall is when such celestials are not heedful;[36] another reason is when humans are unrighteous.

Another discourse extant in both Pāli and Chinese con-

---

[35] EĀ 17.11 at T II 586c28, parallel to AN 4.70 at AN II 75,7 (translated by Bodhi 2012: 458).

[36] AN 5.197 at AN III 243,16 (translated by Bodhi 2012: 815).

firms that unrighteous behavior by the people leads to a lack of rain.[37]

As noted by Harvey (2007: 10), in Buddhist texts "the environment is held to respond to the state of human morality; it is not seen as a neutral stage on which humans merely strut, nor a sterile container unaffected by human actions. This clearly has ecological ramifications: humans cannot ignore the effect of their actions on their environment."

The Ten Courses of Action

The description of moral decline in the Discourse on the World Ruler takes the ten courses of action as its point of reference. Three of these pertain to bodily deeds, four to verbal activities, and three are situated in the mental realm.

The three unwholesome bodily deeds comprise killing, stealing, and sexual misconduct. Refraining from these corresponds to the path factor of right action in the noble eightfold path, a topic to which I will return in the fourth chapter (see below p. 142). The need to abstain from killing, stealing, and sexual misconduct also finds expression in the first three of five precepts incumbent on a Buddhist disciple.

The four verbal deeds among the ten courses of action are false speech, malicious speech, harsh speech, and gossiping. The whole group of four verbal activities corresponds to detailed expositions of wrong speech as opposed to right speech, another factor of the eightfold path. The first of these four is also one of the five precepts.

The remaining three of the ten courses of action belong to the mental realm, comprising greedy desires (or covetousness),

---

[37] AN 3.56 at AN I 160,14 (translated by Bodhi 2012: 254) and its parallel EĀ² 14 at T II 878a15.

ill will, and wrong view. The first of these three stands for the desire to own what belongs to others, the second for wishing that others be harmed or even killed.

Both greedy desires and ill will have a prominent role to play in the Discourse on the World Ruler, where the moral decline sets in when theft occurs. Although stimulated by poverty, stealing is an obvious instance of desiring to own what belongs to others. Ill will in turn becomes gradually more conspicuous until eventually a low point in morality is reached, where human beings are so full of hatred that they just start killing each other.

The last item in the list of ten courses of action is wrong view, which requires further examination. The early discourses describe right view in two different manners (Anālayo 2018a: 30). One of these two speaks of the four noble truths. The other definition of right view concerns eschewing various expressions of mistaken view, such as dismissing the fruitfulness of giving in charity or rejecting a sense of obligation toward one's parents, as well as the denial of a world beyond and of the potential of spiritual practice.

Of particular relevance for environmental concerns is another item mentioned in such definitions, which describes an aspect of wrong view in the following form:[38]

> There are no good and bad deeds, there is no result of good and bad deeds.

In its original setting, this statement refers in particular to the Buddhist doctrine of karma. At the same time, the basic principle described in this formulation can be applied to climate

---

[38] MĀ 15 at T I 437c28, with parallels in AN 10.206 at AN V 293,30 (translated Bodhi 2012: 1536) and Up 4081 at D 4094 *ju* 237b6 or P 5595 *tu* 271b3 (translated by Martini 2012: 64).

change, in the sense of pointing to the need to take responsibility for one's actions.

This is not meant to encourage a sense of guilt. As noted by Klinsky (2019: 475), in particular "conversations about climate injustice can challenge central aspects of people's identities as 'good people'," leading to the need of finding "pathways of opening these discussions that do not feed into denial, retaliation or hardening of protective binary positions."

Here it is also of relevance that the early Buddhist doctrine of karma revolves around intention. It is the intention behind a particular deed that counts. Hence, unintentional harm caused to the environment needs to be clearly distinguished from the same done intentionally.

At the same time, however, once one is aware of the impact of one's actions, the understanding that "there is a result of good and bad deeds" would provide a clear directive for adjusting one's behavior so as to minimize one's carbon footprint. It would follow that, from an early Buddhist perspective, just ignoring the effects of one's actions on the environment conflicts with one of the principles enshrined in canonical definitions of right view.

Ethical Recovery

The need to take responsibility emerges also in the Discourse on the World Ruler, which reports a change of conduct by some human beings that do not participate in the mutual killings:[39]

> Then those who are wise escape far away into [the mountains and] forests and rely on hiding in caves. During those seven days they harbor fear in their hearts. They

---

[39] DĀ 6 at T I 41a29, parallel to DN 26 at DN III 73,7 and MĀ 70 at T I 523b4.

speak [to one another], uttering wholesome words of *mettā*: "Do not harm me; I will not harm you."[40]

By eating grass and the seeds of trees they stay alive. When the seven days are over, they come out of the mountains and forests. Then, on getting to see one another, those who have survived are delighted and congratulate [one another] saying: "You are not dead? You are not dead?"

It is just like parents who have a single son, from whom they have been separated for a long time. On seeing one another, they are delighted without limit.[41] Those people are each delighted like this in their hearts and repeatedly congratulate one another. After that they enquire about their family [and learn] that many of their family members and relatives are dead, so they cry and weep with one another for another seven days. When those seven days are over, they congratulate one another for another seven days, full of joy and delight.

Reflecting on their own [situation], they say: "We accumulated much badness, therefore we encountered this disaster. Our relatives are dead and our family members have disappeared. We should now together cultivate a little what is wholesome. What kind of wholesomeness would it be proper to cultivate? We will not kill sentient beings." At that time [human] beings harbor *mettā* in their hearts,[42] they do not harm one another. Thereupon the appearance and lifespan of these sentient beings increases.

---

[40] DN 26 does not refer to *mettā*.

[41] Whereas DN 26 does not have such a comparison, MĀ 70 at T I 523b7 presents a similar illustration, which here involves just a mother whose only son returns home after a long absence.

[42] The translation is based on adopting a variant reading without a reference to "cessation".

The above passage mentions *mettā*, often translated as "loving kindness", although perhaps better captured with translations like "benevolence" or "goodwill". Such *mettā* finds expression in the words: "do not harm me; I will not harm you." This conveys the sense of a relationship of friendship and protection,[43] cultivated by those who remained established in non-harming.

The nuance of friendship and protection is a regular meaning *mettā* carries in the early discourses. A verse in the Discourse on *Mettā* illustrates this with the example of a mother and her only child, which she would be willing to protect even with her own life.[44] Note that this description is not about a mother's love; in fact, the appropriate Pāli term for such love would be *pema* or *piya* rather than *mettā*. Instead, the verse expresses the sense of the protection a mother would be willing to give to her own offspring (Anālayo 2015b: 29).[45]

Ecological Recovery

The Discourse on the World Ruler continues by depicting how a gradual increase in morality leads to an increase in lifespan. The same also has a beneficial effect on the environment, which gradually recovers. The discourse describes the eventual re-

---

[43] See in more detail, e.g., Collins 1987: 52 and Schmithausen 1997b.

[44] Sn 149.

[45] This sense of protection is directed toward all beings without exception, which implies that it is not merely the expression of an instrumental concern. Though some sentient beings may be of use and others potentially dangerous, this is clearly not the case for "all sentient beings". This would help correct the position taken by Harris 1991: 107 that "there is little evidence in the canon ... to suggest that *mettā* may be extended to other beings simply as an expression of fellow-feeling." For a survey of passages relevant to *mettā* as a simple attitude of kindness and related usages see Maithrimurthi 1999: 48–53.

covery of prosperous living conditions on the earth in the following manner:[46]

> Then this great earth will be open and level, without ra-
> vines, wastelands, or thorny bushes, and there will also
> be no mosquitoes, gadflies, snakes, vipers, or poisonous
> worms. Clay stones, sand, and gravel will become [like]
> lapis lazuli. People will flourish, the five grains will be
> common and cheap, and there will be abundant happi-
> ness without end. Eighty thousand great cities will mani-
> fest, with neighboring villages a cock's crow away from
> one another.

After a period of utter moral and ecological degradation, during which those who wished to survive had to flee and hide in
the wilderness, the peak of recovery expresses itself in a densely
populated earth. Clearly, in this text human civilization as such
is not seen as the problem. The key question is not human dominion over nature but much rather the ethical quality that informs such dominion.

Harvey (2007: 14) comments on the Pāli version of this description that "the message as regards nature, here, seems to be
that, while an immoral society weakens itself and has to look
to nature for renewal, a genuinely moral human civilization
can take over more of the earth without destroying it, perhaps
in an environment of closely clustered human communities—
the text actually says *villages*, towns, and cities—that may still
have nature interspersed within it in semi-wild parks etc."[47]

---

[46] DĀ 6 at T I 41c25, parallel to DN 26 at DN III 75,7 and MĀ 70 at
T I 524b25.

[47] According to Harris 1991: 108, however, "in this perfect world
only urban and suburban environments are left. The jungle has
been fully conquered. Civilization and artifice then are consistent

According to the Discourse on the World Ruler, the earlier deterioration was the outcome of immoral conduct. Conversely, the prosperous living conditions on earth achieved in this way are the result of moral conduct.[48]

Applied to climate change, this suggests that human beings are capable of changing their ways in the face of social and environmental degradation and bring about an improvement of the ecological situation.

The emphasis on moral conduct in this context puts the spotlight squarely on the motivations for unwholesome action. Another discourse explicitly points out where unwholesome conduct originates. The Chinese version of this discourse presents the matter as follows:[49]

---

with the total destruction of the wilderness." Yet, the description need not imply that each village is surrounded on all sides by other villages or towns at very close distance, such that no wilderness at all remains. In view of the existence of areas unfit for village construction, such as rivers, lakes, etc., this would in fact hardly be possible. Instead, the idea could just be that at least one other village will be at a close distance, so that a cock's crow can be heard, leaving open the possibility that in other directions there is no human settlement close by. The point conveyed to the audience could then be that, in order to travel from one city to another (the number eighty thousand is again symbolic and would just convey the idea of "many"), one would be sure of having village after village along the road so that one could easily get provisions, assistance, etc. (see also below p. 103 note 30). On visualizing the resultant spread of human civilization in this way, with cities connected by roads that lead through villages situated in close proximity to each other, there would still be ample space for wilderness in lateral areas.

[48] The future Buddha Maitreya appears only in DĀ 6 and DN 26; he is not mentioned at all in MĀ 70. Moreover, even in DĀ 6 and DN 26 he does not function as a savior; see in more detail Anālayo 2014a.

[49] MĀ 179 at T I 721a3, parallel to MN 78 at MN II 26,12 (translated

Whence do unwholesome [types of] conduct arise? I declare the place from which they arise. One should know that they arise from the mind. What kind of mind? If the mind is with sensual desire, with ill will, or with delusion, one should know that unwholesome [types of] conduct arise from this kind of mind.

The Pāli version takes the same position. In this way, the three root defilements of greed (or sensual desire or sensual craving), hatred (or ill will or anger), and delusion are the motivators for unwholesome conduct. Hence, they are also the chief culprits for the ecological decline described in the Discourse on the World Ruler.

As noted by de Silva (2000: 95), according to the early Buddhist view "the world, including nature and mankind, stands or falls with the type of moral force at work. If immorality grips society, people and nature deteriorate; if morality reigns, the quality of human life and nature improves. Thus greed, hatred, and delusion produce pollution within and without."

Contemplation of the Mind

Recognition of the presence or absence of mental conditions influenced by the three root defilements in one's own mind is a central thrust of the third establishment of mindfulness. The first part of the relevant instructions in a Chinese parallel to the Discourse on the Establishments of Mindfulness (*Satipaṭṭhānasutta*) proceeds in this way:[50]

---

by Ñāṇamoli 1995/2005: 650); for a comparative study of MN 78 see Anālayo 2011b: 424–431.

[50] EĀ 12.1 at T II 568c21, parallel to MN 10 at MN I 59,30 (translated by Ñāṇamoli 1995/2005: 150), see also DN 22 at DN II 299,9 (translated by Walshe 1987: 340), and parallel to MĀ 98 at T I 584a6.

Having a mind with craving for sensual pleasures, monastics are in turn aware of it and know of themselves that they have a mind with craving for sensual pleasures. Having a mind without craving for sensual pleasures, they are also aware of it and know of themselves that they have a mind without craving for sensual pleasures.

Having a mind with anger, they are in turn aware of it and know of themselves that they have a mind with anger. Having a mind without anger, they are also aware of it and know of themselves that they have a mind without anger.

Having a mind with delusion, they are in turn aware of it and know of themselves that they have a mind with delusion. Having a mind without delusion, they are also aware of it and know of themselves that they have a mind without delusion.

The task of mindfulness here is clear recognition of the presence and the absence of these three unwholesome mental conditions. The three covered in the above extract do not exhaust the range of mental states explored in the Discourse on the Establishments of Mindfulness and its parallels, which also cover a liberated or a concentrated condition of the mind, etc. For the present context, however, these first three are particularly relevant.

The attitude behind such mindfulness practice, be it directed toward oneself or others, finds illustration in a simile of a mirror. A Sanskrit version of this illustration proceeds in this way:[51]

---

[51] Gnoli 1978: 248,26, parallel to DN 2 at DN I 80,15 (translated by Walshe 1987: 106) and DĀ 27 at T I 109b8 (to be supplemented from DĀ 20 at T I 86a28, as DĀ 27 abbreviates; for a translation of this part see Anālayo 2019g: 1928). The simile is not found in another two parallels, EĀ 43.7 and T 22.

It is like a clear-sighted person who has taken hold of a round mirror that is very clear and were to examine the image of one's own face.

A Chinese parallel just mentions looking into clear water to see oneself. The Pāli version has both options, either a mirror or a bowl with clear water. Whichever tool is used, the task is to recognize clearly whether a state like anger, for example, arises.

Internal and External Mindfulness

The illustration of looking into clear water or a mirror applies not only to one's own mind. In fact, in its original setting the simile illustrates recognition of the mind state of another.

Already in the previous chapter the need to proceed from what is internal to what is external became evident in relation to the earth element. The same pattern holds for the mind, in that there is progression from the internal to the external. Such a progression finds explicit description in another passage, which proceeds as follows:[52]

> One contemplates the body internally with untiring energy and collected mindfulness that is not lost, removing greed and sadness in the world, and one contemplates the body externally with untiring energy and collected mindfulness that is not lost, removing greed and sadness in the world. Contemplation of feeling tones, the mind, and dharmas is also like that, with untiring energy and collected mindfulness that is not lost, removing greed and sadness in the world.
>
> Having contemplated the body internally, one arouses knowledge of the bodies of others. Having contemplated feeling tones internally, one arouses knowledge of the feeling tones of others. Having contemplated the mind internally, one arouses knowledge of the minds of others. Having contemplated dharmas internally, one arouses knowledge of the dharmas of others.

---

[52] DĀ 4 at T I 35c27 parallel to DN 18 at DN II 216,10 (translated by Walshe 1987: 298). Another parallel, T 9 at T I 216a8, does not proceed beyond listing internal and external contemplation. I have explored internal and external mindfulness practice in more detail in Anālayo 2003: 94–99, 2013b: 17–19, and 2020c.

From this perspective, mindful observation of one's own mind as "internal" contemplation forms the starting point for then trying to discern the mental condition of others. Contemplation of the mind undertaken in this way could be applied to environmental activity as a way of monitoring one's own mental condition and that of others in order to ensure that any activism will be undertaken in a way that offers the best chance of success.

Nhat Hanh (2008: 18) even frames the whole situation from the viewpoint of mindfulness: "the situation the Earth is in today has been created by unmindful production and unmindful consumption." By now, "the bells of mindfulness are sounding. All over the Earth, we are experiencing floods, droughts, and massive wildfires. Sea ice is melting in the Arctic and hurricanes and heat waves are killing thousands. The forests are fast disappearing, the deserts are growing, species are becoming extinct every day, and yet we continue to consume, ignoring the ringing bells" (2008: 1). Nhat Hanh (2008: 77) then sums up that "truly engaged Buddhism is first of all practicing mindfulness in all that we do."

Thanissara (2015: 138f) reasons that "while there are a lot of things we can 'do', it's also important to explore the quality of awareness that informs our 'doing'. A shift from an oil-based economy to renewable energy is going to be a long struggle, so we need to pace ourselves and avoid getting caught up in overwhelm, defeatism, anger, and burnout. For this an inner practice is essential."

Strain (2016: 201) adds that the "practice involves self-transformation as well as action upon the world." From that perspective, according to Curtin (2017: 30) "every problem, if we dig down deeply enough, is an invitation to practice."

As pointed out by Bodhi (2009: 162): "effective action must

be rooted in insight, in wisdom." Ringu (2009: 133) reasons that "the worse the situation in the world gets, the stronger our spiritual practice could become. External chaos becomes an eye-opener. It demonstrates the futility of chasing worldly things … whatever problems appear, on the outside or the inside, whether we are sick or dying, even if everything works out in the worst way, we must maintain internal harmony, clarity, and compassion. We relax and do what we can. We learn not to be overpowered by events and general human behavior."

In this way, the overall suggestion would be to view the current dilemma from the perspective of internal and external mental states and their ethical quality. Those responsible for preventing the necessary actions being taken and for spreading misinformation about climate change are under the influence of defilements. Rather than getting upset with individuals, the whole situation can be evaluated under the overarching concern of countering the detrimental influence of defilements, within and without.

In line with the general pattern of proceeding from the internal to the external, the starting point would be first of all discerning even subtle traces of the three root defilements in one's own mind. This can take place by matching greed, hatred, and delusion with three possible responses to the current crisis, namely denial, anger, and resignation. In what follows I explore these three responses, in the understanding that the relationships proposed here to the three root defilements are merely my own suggestion and not something found in this way in early Buddhist thought.

Denial

When facing information about ecological destruction and climate change with their potential repercussions, it is natural to want to avoid and ignore it. One wishes to continue to enjoy the pleasures of this world without having to be too concerned about the consequences. In this way denial, which can be considered an expression of the root defilement of greed, prevents reacting appropriately to what is taking place.

The forces of greed are strong enough to make denial an intentional strategy by some leading politicians and high-level business executives who would be affected by actions taken to counter the crisis (Almiron and Xifra 2020).

A common mode of such denial is to pretend that the information we have is not sufficiently well-established to be taken seriously. Yet, for human beings to respond it is in principle enough to know that a threat is probable; there is no need to be completely certain. This is part of how human perception works, which involves "perceptual prediction" (Anālayo 2019d).

On suddenly encountering a dangerous animal, one will react on the spot. In such a situation one cannot afford to wait until all the information about the animal has been gathered and one is completely sure that it intends to attack, since by then it may be too late. Similarly, faced by the potential outcomes of the current climate crisis, it is time to act now, before it is too late.

The tendency to want to ignore can exert a strong influence that is hardly noticed, unless mindfulness is established. From this viewpoint, the ecological crisis can become an opportunity for regular scrutiny of the mind in order to detect the potential influence of the root defilement of greed, however subtly it might manifest, in fostering denial. The task is to keep bringing the mind back on track; in the words of Goldstein (2009: 182), "we need to repeatedly remind ourselves of the situation and not settle for a generalized understanding that climate change is a problem."

Anger

Another type of reaction to the ecological crisis is anger. Given that some are actively working to prevent appropriate changes from taking place, it might seem natural to get angry with them. Yet, this is not a solution. For one, to some degree most human beings contribute to the problem. Let the one who has never driven a car, taken a flight, eaten food imported from abroad, worn clothing manufactured in a distant country, etc., throw the first stone.

Pointing out that nearly everyone is part of the problem is not meant to inculcate a sense of guilt. The question is not to turn one's carbon footprint into some sort of original sin and then feel a need to atone for that. The task is only to take responsibility and act in a conscientious manner, without shifting all the blame onto others.

From an early Buddhist perspective, even righteous anger is a defilement of the mind, a topic I will explore again in the next chapter (see below p. 91). There is definitely a place for stern and strong action, but this should always come with inner balance rather than aversion. Inner balance is crucial for any possible environmental activity to achieve maximum benefit. From the viewpoint of mindfulness practice, getting angry is succumbing to one of the root defilements and thereby to what has contributed to and sustains this very crisis. Anger is a problem and not a solution. A solution can only be found when the mind is not clouded by defilements and therefore able to know and see things accurately.

Although anger puts things into sharp relief and thereby can create a superficial impression of clarity, it actually distorts perceptual appraisal of the situation and prevents a balanced and correct discernment. A mind in the grip of anger is unable to see accurately what is for one's own benefit and for the benefit of others. Anger encourages a tunnel vision that ignores aspects of the situation that do not accord with its evaluative thrust.

Such a condition of the mind needs to be recognized with mindfulness; it calls for refraining from taking action until the mind has cooled down, as only then does it become possible to see things in their proper perspective. In this way, the current crisis is best handled with a mind that is not in a condition of anger.

Instead of anger, compassion is the appropriate motivating force for taking action. Its cultivation can provide all the energy needed to become active to correct wrongs, rather than just letting injustices persist.

In order to counter anger or other negative mental conditions, it can at times be helpful to work against a black-and-white perception of the situation by reminding oneself that in the course of evolution greenhouse gas been responsible for enabling life on the planet, as without its effect the earth would be too cold for life. Furthermore, human use of fossil fuels has in the past facilitated some beneficial developments.[53]

Of course, now the time has come to change, as the continued overuse of fossil fuels has turned what was a protective function of greenhouse gas into something highly dangerous. Due to the influence of defilements, narrow mindedness, and selfishness, some have not yet realized that it is high time to adjust. No need to hate them for that. Instead, out of a compassionate sense of responsibility for life on this planet, the need-

---

[53] Wallace-Wells 2019: 53 notes that "the graphs that show so much recent progress in the developing world—on poverty, on hunger, on education and infant mortality and life expectancy and gender relations and more—are, practically speaking, the same graphs that trace the dramatic rise in global carbon emissions."

ful should be done to promote the necessary change. This can involve stern and strong actions but at the same time remain rooted in a mental attitude of compassion and equanimity.

Resignation

The third reaction to the environmental crisis to be discussed here is resignation. This could be related to the root defilement of delusion. Needless to say, delusion of course also underlies the previously surveyed reactions of denial and anger. Relating the present type of reaction to delusion is therefore only meant in the sense that such resignation is not evidently influenced by either greed or aversion, and at the same time involves some degree of lack of clarity and proper perspective.

Resignation can easily manifest in a sense of feeling overwhelmed and helpless. As a single individual, it just seems hopeless to try to effect any change. What is the point of even trying?

Yet, society is made up of individuals and does not exist apart from them. The question is not whether a single individual can bring about all required change alone. The question is rather whether every single individual can contribute to the required change. This is indeed the case.

As noted by Kaza (2014: 86), "Buddhist ethics view the individual as an active agent in a vast web of relationships where every action generates effects. Based on this worldview, I would argue that attaining ecological and economic sustainability under the challenges of rapid climate change requires ethical engagement."

This understanding can be employed to counter the assumption of mono-causality, be it consciously or unconsciously, in the sense that just a single cause is held to be responsible for a particular situation or problem. Such an assumption can easily lead to searching for a single culprit that can serve as the scapegoat for one's negativities. It can also result in over-estimating one's own personal responsibility and as a result falling prey to sentiments of helplessness in view of the magnitude of the problem. Viewing oneself and others instead as co-participants in a large network of conditions can serve to counterbalance such tendencies.

Another relevant teaching is right view. As mentioned above, one aspect of right view is the recognition that good and bad actions produce results. In relation to environmental destruction, this points to taking responsibility for one's actions. Dzigar Kongtrul (2009: 147f) reasons that, out of a sense of responsibility, "we have to address global warming ... thinking negatively, blaming others, or feeling hopeless will not help ... taking action will bring transformation."

Summary

Early Buddhist thought posits a close relationship between moral conduct and environmental conditions, to the extent of attributing a serious decline in living conditions, in some respects similar to the anticipated results of climate change, to a lack of ethical restraint.

Ethical conduct in turn reflects the condition of the mind. Of central importance here are the three root defilements of greed, hatred, and delusion. A focus on countering the impact of these three root defilements can serve as an expression of an environmental concern and lead to corresponding action. On the internal level, these can manifest in the form of denial, anger, and resignation. Cultivating mindfulness to face and overcome these conditions can serve as a training ground for confronting their external manifestations, in particular when these have repercussions on the environment.

Informed by the early Buddhist teachings on conditionality and right view, reliance on mindfulness facilitates stepping out of unhelpful attitudes in order to make one's individual contribution, to the best of one's abilities, toward protecting living conditions on this earth.

Liberation of the Mind

> What is the truth of the cessation of *dukkha*? It is being able to bring about the eradication and cessation without remainder of that craving, so that it will not arise again. This is reckoned to be the truth of the cessation of *dukkha*.[1]

The third truth envisages nothing short of complete liberation of the mind from all craving and defilements and therewith from each of the three root poison of greed, hatred, and delusion.

Such complete liberation is the overarching concern of early Buddhist thought, the pinnacle toward which all other teachings point in one way or another. In order to do justice to this orientation, the approach to facing climate change I present here needs to be related in some way to progress to full awakening, the highest level of liberation recognized in early Buddhism. I will take up this topic in more detail at the outset of the next chapter (see below p. 119). In this chapter, I focus on the divine abodes as types of liberation that are in themselves temporary in nature but which at the same time are natural conditions of the mind of one who has walked the path to full liberation.

After a preliminary survey of the four levels of awakening, I turn to passages taken up in previous chapters that point to the relevance of the four divine abodes. Then I examine the simile of the saw and the question of whether the Buddha, after having reached awakening, was still subject to anger. Next, I explore communal harmony as an expression of *mettā*, the

---

[1] EĀ 24.5 at T II 619a15.

opposite of anger, followed by studying a simile on acrobats performing together as an illustration of how to protect oneself and others through *mettā* and mindfulness.

Out of the four divine abodes, compassion is particularly relevant to environmental concerns and for this reason is a central topic in the present chapter. I begin by exploring compassion directed to non-human sentient beings. Based on ascertaining the basic thrust of compassion as the wish for the absence of harm, I then contrast compassion to grief and set it within the framework of the eightfold path. After that, I explore the ethical dimension of the divine abodes in general. In the final part of this chapter I take up the boundless radiation of the divine abodes and illustrate their meditative experience.

Types of Liberation

Early Buddhist thought recognizes different types of liberation (Anālayo 2009b). Of central importance are liberations of the definite type, attained with the four levels of awakening. According to the early discourses, these four levels involve the eradication of certain fetters and have irreversible consequences for the future.

With the first level of awakening, three out of a standard set of ten fetters have been eradicated. As a result, one becomes a stream-enterer who is sure to reach full awakening within a period not exceeding seven more lifetimes. The three fetters left behind for good are belief in the existence of a permanent self, doubt (of the fundamental type regarding the teachings), and clinging to rules and observances as in themselves leading to liberation.

With the second level of awakening a practitioner becomes a once-returner, having substantially reduced the next two fetters of greed and anger, and being certain to reach full awakening within one more lifetime.

Progress to the third level of awakening leads to becoming a non-returner, in the sense of no longer returning to be born in the world of humans and lower celestial beings. Having successfully overcome five out of the set of ten fetters, the rebirth of a non-returner can only take place in a superior heavenly realm inhabited solely by those who are completely free of any greed and anger.

The remaining five fetters to be left behind with the attainment of full awakening, the highest of the four levels of liberation recognized in early Buddhist thought, are desire for fine-material and for immaterial becoming, conceit, restlessness, and ignorance. An arahant has actualized the full liberating potential of the teachings by eradicating all ten fetters and becoming free from the prospect of future rebirth in *saṃsāra*.

Alongside these levels of awakening expressing the core values and orientations of early Buddhist soteriology, other liberations recognized in the early discourses are of a more temporary type. Such liberations of the mind result from the gaining of mental tranquility, when defilements have gone into abeyance for the time being. Although such experiences have considerable transformative potential, they are in themselves deemed unable to achieve the eradication of the fetters described above (Anālayo 2017b: 184–192 and 2020a).

Among such temporary liberations, of particular relevance to my present exploration are the four "divine abodes" (*brahmavihāra*). In a way, these four can be considered forms of abiding in a celestial condition while still being a human on earth. Alternatively, they are also referred to as states that are "boundless" or "immeasurable" (*appamāṇa*). The mind of one who dwells in any of these four has become boundless and immeasurable.

The four boundless states or divine abodes comprise *mettā*, already discussed above (see p. 69), compassion, sympathetic joy, and equanimity. Out of the whole set of four divine abodes, in the early discourses *mettā* features with particular prominence (Anālayo 2015b: 28–39). As a basic mental disposition of kindness, well-wishing, and mutual protection, *mettā* is relevant to the whole gamut of bodily, verbal, and mental conduct, ranging from daily activities to formal meditation.

Building on the foundation laid in this way, compassion takes its place as a mental disposition of aspiring for the absence of harm and affliction. Compassion is not just *mettā* toward those in distress (Anālayo 2019c). Both *mettā* and compassion can be directed toward all sentient beings. It follows that the difference between them cannot be merely due to the object they take.

Compassion has a complement in sympathetic joy, the intentional cultivation of rejoicing in the manifestation of wholesome qualities, being directly opposed to envy and discontent.

The cultivation of the divine abodes culminates in equanimity. Such equanimity is the very opposite of indifference. It is an inner stability that comes with an open heart, matured through previous development of the other three divine abodes.

These four attitudes are characteristic of the mind of an arahant, who by dint of having eradicated all defilements will respond to whatever happens in one of these four modes. From this perspective, progress through the four levels of awakening, described above, could be visualized as a gradual opening of the heart until, with full awakening attained, it completely embodies the divine abodes.

The Role of the Divine Abodes

The relevance of the divine abodes to the topics explored in this book can be illustrated with some of the passages surveyed in earlier chapters. The Chinese version of the Discourse on the World Ruler, for example, brings up *mettā* at the key turning point of its narrative. Once ethical and environmental decline have reached an all-time low and human beings randomly kill each other, those who have gone into hiding establish a foundation for a gradual improvement of ethics and consequently of the environment in the following manner:[2]

> They speak [to one another], uttering wholesome words of *mettā*: "Do not harm me; I will not harm you."

The Pāli version does not explicitly mention *mettā* at this juncture. All versions agree, however, that the survivors of the massacre decide to refrain from killing, which implies the same principle. A description of the first precept of abstaining from killing provides an explicit relationship to the divine abodes, which takes the following form:[3]

> I abstained from killing and have abandoned killing, discarding sword and club, having a sense of shame and fear of blame, with a mental attitude of *mettā* and compassion for the welfare of all [sentient beings], even insects.

It follows that *mettā* and compassion are the key qualities which in the Discourse on the World Ruler lead to the decisive shift toward an improvement of ethics and a gradual recovery of the environment. The same qualities could play a similar role in relation to the present ecological crisis.

---

[2] DĀ 6 at T I 41b1, with a parallel in MĀ 70 at T I 523b7.

[3] MĀ 187 at T I 733a28, with a parallel in MN 112 at MN III 33,19 (translated by Ñāṇamoli 1995/2005: 907, to be supplemented from p. 448). A minor difference here is that MN 112 does not explicitly mention *mettā* or the killing of insects.

The whole set of four divine abodes features in another discourse, taken up in the first chapter, which presents the earth as an inspiration for cultivating mental balance (see above p. 33). The relevant instruction is to remain balanced with whatever happens, however disagreeable it might be, based on the injunction:[4]

You should maintain your mind like the earth.

The exposition proceeds by taking up the other elements and then the four divine abodes. The Pāli version provides additional specifications on the purpose of each of these four,[5] indicating that *mettā* counters ill will, compassion cruelty, sympathetic joy discontent, and equanimity (lustful desire) and aversion.[6]

These indications reflect the potential of the divine abodes to make a substantial contribution to progress to awakening. I will return to this aspect of the divine abodes below, in particular in relation to compassion. Suffice it for now to note that, just as progress to awakening can lead to the flowering of the divine abodes, so can a nurturing of these four sublime attitudes substantially empower the cultivation of insight aimed at awakening.

---

[4] EĀ 43.5 at T II 760a5 and MN 62 at MN I 423,18 (translated by Ñāṇamoli 1995/2005: 529).

[5] MN 62 at MN I 424,27. EĀ 43.5 at T II 760a11 recommends the arousing of the four divine abodes but then proceeds on to another topic, which could be the result of a shifting of textual portions that appears to have occurred in EĀ 43.5; see Anālayo 2014/2015.

[6] The supplementation in brackets is motivated by the fact that, although MN 62 at MN I 424,33 contrasts equanimity just to aversion, *paṭigha*, DN 33 at DN III 249,14 contrasts the same divine abode to greed or lust, *rāga*. From a practical perspective, equanimity would indeed counter both,

The Simile of the Saw

Another instance reflecting the relevance of the divine abodes occurs in the Discourse on the Elephant's Footprint, taken up in the first chapter, whose exposition leads up to the powerful simile of the saw. To recapitulate, the Discourse on the Elephant's Footprint first relates the internal to the external earth element. Then it shows how insight cultivated in this way fortifies one when experiencing abuse.

The same insight can inspire the cultivation of equanimity to such an extent that one will be able to maintain inner balance even when being physically attacked. It is at this juncture that a reference to the well-known simile of the saw occurs:[7]

> Friends, the Blessed One has also spoken in this way: "Suppose there are bandits who come and cut your body apart, limb by limb, with the help of a sharp saw. If, when those bandits cut your body apart, limb by limb, with the help of a sharp saw, there is an alteration in your mind, or you speak bad words, you are regressing.
>
> "You should have this thought: 'If there are bandits who come and cut my body apart, limb by limb, with the help of a sharp saw, because of that there will not be an alteration in my mind and I will not speak bad words. I will arouse a mind of compassion toward those who are cutting my body apart limb by limb.'"

The original delivery of this simile occurs in the Discourse on the Simile of the Saw, where it leads on to the practice of *mettā*.[8]

---

[7] MĀ 30 at T I 465a6, parallel to MN 28 at MN I 186,10 (translated by Ñāṇamoli 1995/2005: 279).

[8] MN 21 at MN I 129,15 (translated by Ñāṇamoli 1995/2005: 223) and its parallel MĀ 193 at T I 746a12; for a comparative study of MN 21 see Anālayo 2011b: 145–147.

The dramatic depiction in this passage is probably best understood as referring to a situation where one is completely at the mercy of bandits, without any possibility of preventing their cruel deed (Anālayo 2019c). In other words, it can safely be assumed that the description is not meant to discourage one from trying one's best to avoid being cut up by bandits or undergoing any other type of abuse. The simile appears to be only intended to illustrate, with the help of a rather drastic situation, that even under the most excruciating circumstances a carrying out of the teachings of the Buddha requires not reacting with anger and abuse.

This in turn provides strong support for the suggestion made in the previous chapter that an implementation of the early Buddhist teachings in the context of ecological or any other social activism leaves little scope for endorsing anger, however righteous it may appear.

The Buddha and Anger

In view of the powerful teaching on the simile of the saw, it would be natural to expect that the Buddha himself set an example for being free of anger. As mentioned at the outset of this chapter, a condition characteristic of all those who have reached full awakening, Buddhas and arahants, is complete freedom from the influence of the three root defilements.[9]

An impression that at first sight might seem to conflict with this normative view is a famous episode involving a monastic and relative of the Buddha by the name of Devadatta, who ac-

---

[9] DN 29 at DN III 133,21 (translated by Walshe 1987: 435) and its parallel DĀ 17 at T I 75b18; see also Sanskrit fragment Or.15009/565r4, Nagashima 2015: 394, which appears to have preserved a reference to the earlier-mentioned impossibility of intentionally killing a sentient being.

cording to the textual sources was the first in the history of Buddhism to create a schism in the monastic community.

Several monastic codes report that Devadatta openly asked the Buddha to hand over leadership of the community to him. In reply, the Buddha is on record for having refused, comparing Devadatta to spittle.[10] The employment of such terminology could easily convey the impression that the Buddha got angry.[11]

When examining this episode, the function of narratives in texts on monastic discipline needs to be taken into account. The overarching concern of such narratives is not the accurate reporting of historical details. Instead, they often serve a didactic purpose, where the effect of a particular story on the audience is decisive.[12] In order to detect the influence of such concerns, comparative study can often be of considerable help.

In the present case, the relevant parts in a text on monastic discipline of a Buddhist tradition that developed independently from an early date in the history of Buddhism, the Mahāsāṅghika *Vinaya*, has no record of this episode at all.[13] A discourse extant in Chinese, found in the *Ekottarika-āgama*, also does not report any request by Devadatta for the leadership of the

---

[10] The Dharmaguptaka *Vinaya*, T 1428 at T XXII 592b13, the Mahīśāsaka *Vinaya*, T 1421 at T XXII 18b20, the Mūlasarvāstivāda *Vinaya*, T 1442 at T XXIII 701c21, the Sarvāstivāda *Vinaya*, T 1435 at T XXIII 258b7, and the Theravāda *Vinaya*, Vin II 188,37 (translated by Horner 1952/1975: 264); for further sources see Lamotte 1970b: 1671–1673.

[11] For example, Levman 2019: 32 speaks of a "very real and very un-Buddhistic human reaction of the Buddha to Devadatta's machinations, that is, his apparent anger." His assessment comes without a reference and could also be meant to refer to a prediction attributed to the Buddha that Devadatta will go to hell.

[12] See in more detail Anālayo 2012a, 2014b, and 2016e.

[13] T 1425 at T XXII 281c12 and 442c29.

monastic community.[14] This discourse shares several other episodes in common with those texts on monastic discipline that report the Buddha comparing Devadatta to spittle.

From a comparative perspective, it seems highly probable that this particular story is a later addition to the repertoire of stories about Devadatta.[15] Hence, the depiction of the Buddha's use of offensive language quite probably reflects the sentiments of the narrators of this episode; it is not an indicator that, according to early Buddhist thought, fully awakened ones still get angry.

The Pāli discourses recurrently depict the Buddha using the term *moghapurisa* when rebuking disciples who have misrepresented him or behaved in an obstinate manner. Occurrences of the first part of this term in a negative sense (*amogha*) convey that something is useful.[16] On adopting this meaning, the term *moghapurisa* could be rendered as a "useless person". On this understanding, its employment need not be taken to convey that the Buddha got angry.[17] Instead, it can simply be con-

---

[14] EĀ 49.9 at T II 802b15.

[15] Mukherjee 1966: 144–147, Bareau 1991: 102f, Ray 1994: 168f, Deeg 1999: 186, and Li 2019: 141–143; see also Lamotte 1970a and Jing 2009.

[16] Examples are DN 19 at DN II 252,12: *amoghā pabbajjā*, "their going forth was not in vain," SN 9.14 at SN I 232,26: *amogham tassa jīvitam*, "one's life was not in vain," AN 1.6 at AN I 10,21: *amogham raṭṭhapiṇḍam bhuñjati*, "one partakes of the country's alms food not in vain." Notably, the last statement concerns a monastic who just for a short moment cultivates *mettā*, which shows the importance accorded to such practice.

[17] Levman 2019: 42 comments on occurrences of the term *moghapurisa* in reproaches attributed to the Buddha that this "seems unnecessarily harsh and condemnatory", reasoning that the early texts show the Buddha to have been "reproachful, even angry".

sidered a strong form of admonishment that is often warranted by the narrative context.

In sum, the early texts do show the Buddha at times being rather stern and openly expressing censure, but this need not be taken to imply that he got angry. Instead, such descriptions are best read as depicting verbal activity that springs from a mental attitude of equanimity, grounded in a mind that is firmly established in what is wholesome.

Communal Harmony

As the exact opposite of anger and ill will, the first divine abode of *mettā* is a central quality to ensure communal harmony. This function has several dimensions. A passage describing such communal harmony begins by depicting a basic attitude to be aroused toward others:[18]

> I think: 'It is to my good benefit, it is a great fortune for me, namely, that I am together with such practitioners of the holy life.'

The reference to the "holy life" here is an idiom in the early discourses to express that the protagonists of this discourse were living the life of Buddhist monastics. The basic appreciation of one's companions, described in this passage, forms a foundation that leads on to the cultivation of *mettā* in the following manner:[19]

---

[18] MĀ 185 at T I 730a6, parallel to MN 31 at MN I 206,19 and a Gāndhārī fragment, Silverlock 2015: 369. For a comparative study of MN 31 that also takes into account EĀ 24.8 see Anālayo 2011b: 203–209; see also Allon and Silverlock 2017: 16f

[19] MĀ 185 at T I 730a18, parallel to MN 31 at MN I 206,21 and Silverlock 2015: 370, which do not have a reference to cultivating such *mettā* "equally and without discrimination".

I constantly practice towards these practitioners in the holy life bodily acts of *mettā*, overtly and covertly, equally and without discrimination. I practice [towards them] verbal acts of *mettā* and I practice mental acts of *mettā*, overtly and covertly, equally and without discrimination.

This description shows the broad range of applicability of *mettā* in daily life, where its cultivation can encompass the bodily, verbal, and mental domain. Moreover, it can take overt forms, which are evident to others, but it can also be done in ways that are not noticed by others.[20] The pervasive cultivation of *mettā* in all these ways then leads up to the following reflection:[21]

I think: 'I would now rather let go of my own ideas and follow the ideas of those friends.' Then I let go of my own ideas and follow the ideas of those friends. We never have a single discordant idea.

On hearing this report, the Buddha concluded:[22]

---

[20] As an example for covert (literally "not seen" in MĀ 185) bodily deeds of *mettā*, the commentary, Ps II 239,23 (translated by Aronson 1980/1986: 33), describes seeing something accidently left in disorder by another. Without further comment, one just quietly sets it in order.

[21] MĀ 185 at T I 730a21, parallel to MN 31 at MN I 206,25 and Silverlock 2015: 370, which do not have a counterpart to the reference to a single discordant idea and instead mention that, although having different bodies, they were as if they had a single mind.

[22] MĀ 185 at T I 730a25, parallel to MN 31 at MN I 206,12, where a comparable remark occurs earlier, as part of the inquiry by the Buddha that then motivates the monastics to report how they were dwelling together in harmony. Another minor difference is that MN 31 does not emphasize that they had a single teacher and

In this way you are constantly together in harmony, at ease and without dissention, of a single mind, of a single teacher, blending like water and milk.

The description offered in this way illustrates different dimensions of living together in communal harmony, which take off from the basic mental disposition of *mettā*. Such communal harmony provides a direct contrast to the depiction of mutual hatred and killing in the Discourse on the World Ruler.

Applied to the current situation, if these principles of communal harmony and *mettā* had become more prominent as guiding principles at some time in the past, the dire repercussions of climate change would not be happening in the first place.

The Acrobat Simile

Another teaching relevant to the divine abodes and human relationships is a simile that involves two acrobats who are about to perform a feat together.

One acrobat tells the other that they should make sure to take care of each other, so that they can perform well. The other responds that this will not do. They first of all need to take care of themselves. It is in this way that they will be able to protect both themselves and the other, and at the same time perform well.

A Chinese version of this discourse draws out the implication of this simile in the following way:[23]

[How does protecting oneself protect others]? Becoming

---

instead notes that they looked on each other with affectionate eyes.

[23] SĀ 619 at T II 173b14 (translated by Anālayo 2012d: 2), parallel to SN 47.19 at SN V 169,15 (translated by Bodhi 2000: 1648) and T 1448 at T XXIV 32b22 (translated by Anālayo 2019g: 1934).

familiar with one's own mind, cultivating it, protecting it accordingly, and attaining realization; this is called 'protecting oneself protects others.'

How does protecting others protect oneself? By the gift of fearlessness, the gift of non-violation, the gift of harmlessness, by having a mind of *mettā* and empathy for others; this is called 'protecting others protects oneself.'

For this reason, monastics, you should train yourself like this: 'Protecting myself, I will cultivate the four establishments of mindfulness; protecting others, I will also cultivate the four establishments of mindfulness.'

In order to be able to perform well, first the two acrobats need to make sure that they are each centered themselves. Based on having in such a way protected their own balance, they will be able to protect each other.

It is noteworthy that the key element here is mindfulness. The arousal of mindfulness builds the foundation for being able to protect oneself through cultivation of the mind and protect others in the form of being compassionate toward them, avoiding any harm.

Self-protection through mindfulness is what enables recognition of the presence of greed, hatred, and delusion, be it within or without. In the absence of such recognition, the three root defilements will have free rein to wreak havoc in the mind, hiding under the various pretenses and excuses that can serve to disguise their true nature. Mindfulness enables seeing through these disguises.

In this way, established mindfulness can counteract an innate unwillingness to admit to oneself that one is greedy, angry, or confused. To the degree to which mindfulness becomes a continuous self-protection, to that same degree it also protects others from the repercussions of any action influenced by the three root defilements.

As explained by Ñāṇaponika (1990: 5), "moral self-protection will safeguard others, individuals and society, against our own unrestrained passions and selfish impulses ... they will be safe from our reckless greed for possessions and power, from our unrestrained lust and sensuality, from our envy and jealousy; safe from the disruptive consequences of our hate and enmity, which may be destructive or even murderous; safe from the outbursts of our anger and from the resultant atmosphere of antagonism and conflict which may make life unbearable for them."

Based on such self-protection, *mettā* and compassion naturally fall into place. Their very cultivation, by dint of taking place in one's own mind, will have immediate beneficial effects within and at the same time contribute to protecting oneself. A mind that dwells in *mettā* and compassion is far removed from intentions of harm, from getting irritated and annoyed, whereby others are protected.

Compassion for Sentient Beings

Nhat Hanh (2008: 72) applies the idea of mutual protection to environmental concerns in the following manner: "to protect the non-human elements is to protect humans, and to protect humans is to protect non-human elements."

The notion of affording protection resonates in particular with compassion. Such compassion can be illustrated with the help of an episode reported in the Pāli *Vinaya*.[24] The relevant section reports that a monastic had released an animal that had been caught in a hunter's trap.

---

[24] Vin III 62,30 (translated by Horner 1938/1982: 105).

Afterwards the monastic became worried, wondering if by acting in this way he might have violated the precept against theft. When he related what he had done to the Buddha, the latter inquired after the monastic's motivation. When the monastic clarified that he had acted out of compassion, the Buddha concluded that for this reason it did not involve a breach of the monastic code.

The *Vinaya* continues with cases where the same action is not motivated by compassion but with a mind intent on theft, in which case the Buddha ruled that it constituted an offence. This is no minor matter, since theft of something valuable counts among the most serious breaches in monastic law, resulting in irreversible loss of one's status as a fully ordained monastic (Anālayo 2016c and 2019f).

The fact that the motivation of compassion for the suffering animal makes the crucial difference is telling. It shows the importance accorded in this particular monastic text to compassion for animals.

Compassionate concern for other sentient beings and the wish to avoid that harm befalls them is a recurrent feature in the early texts. An example is the Buddha's instruction to a lay supporter that leftover food should be discarded in such a way that it does not result in pollution that could harm sentient beings.[25] The same circumspect behavior occurs as part of the description of the harmonious cohabitation of a group of monastics, already taken up above. The monastic who took his meal last would clean their communal eating area in the following way:[26]

---

[25] SN 7.9 at SN I 169,2 (translated by Bodhi 2000: 263), with parallels in SĀ 1184 at T II 320c20 and SĀ² 99 at T II 409b6; see also the discussion in Schmithausen 1991: 30f.

[26] MĀ 185 at T I 729c11, parallel to MN 31 at MN I 207,16 (translat-

If there were leftovers, he would pour them out on to a clear piece of ground or into water that contained no sentient beings.

Such passages reflect a need, relevant to both laity and monastics, to avoid pollution, which as an expression of compassion could be applied to the current ecological situation.

Compassion as Non-Harm

In the early discourses in general, compassion stands for the absence of cruelty and harming. This finds explicit expression in a discourse which identifies what unwholesome mental condition directly opposes compassion. The passage proceeds in this way:[27]

Suppose a monastic says: "I practice liberation [of the mind] by compassion and cruelty arises in my mind." Another monastic [should] tell him: "Do not make this statement, do not slander the Tathāgata; the Tathāgata does not make such a statement. That cruelty still arises in those who with dedication apply themselves to the cultivation of liberation [of the mind] by compassion, that is impossible."

---

ed by Ñāṇamoli 1995/2005: 302). Another parallel to this description, EĀ 24.8, does not mention in what way leftover food was discarded; the same holds for what has been preserved in the Gāndhārī fragment parallel, Silverlock 2015: 369.

[27] DĀ 10 at T I 54b6 (text has been supplemented from DĀ 10 at T I 54b3, as the original abbreviates the exposition of compassion), parallel to DN 34 at DN III 280,27 (DN 34 abbreviates and hence needs to be supplemented from DN 33 at DN III 248,12, translated by Walshe 1987: 500), Sanskrit fragments, Mittal 1957: 78 (§VI.7.2), and T 13 at T I 236a8.

The term Tathāgata here refers to the Buddha.[28] The passage shows that compassion is the very opposite of cruelty and harming.

Compassion as the aspiration for the absence of harm, which is of course of considerable relevance for environmental concerns, can be illustrated with the help of another passage. The passage occurs in a description of how to avoid resentment toward someone who behaves in thoroughly unwholesome ways. The recommendation is that one should try to find something good in this badly-behaving person and pay attention to that. However, if nothing good at all can be found, then such a person provides an excellent opportunity for the cultivation of compassion.

This is the appropriate response here, in that the unwholesome conduct of others is no excuse for getting angry. The passage supports the observation made earlier that, from an early Buddhist perspective, anger is not an appropriate response. This holds even when another is behaving in ways that are entirely irresponsible and reprehensible.

The cultivation of compassion then finds the following illustration:[29]

> It is like a person who is on an extended journey along a long road. Becoming sick halfway, he suffers extremely and is exhausted. He is alone and without a companion. The village behind is far away and he has not yet reached the village ahead. Suppose a person comes and, standing to one side, sees that this traveler, who is on an extended journey along a long road, has become sick halfway, suf-

---

[28] On different nuances of the term see Anālayo 2017e.
[29] MĀ 25 at T I 454b18, parallel to AN 5.162 at AN III 189,8 (translated by Bodhi 2012: 776).

fers extremely, and is exhausted. He is alone and without a companion. The village behind is far away and he has not yet reached the village ahead.[30]

[The second person thinks:] "If he were to get an attendant, emerge from being in the far away wilderness and reach a village or town, and were to be given excellent medicine and be fed with nourishing and delicious food, be well cared for, then in this way this person's sickness would certainly subside."

That is, that person has extremely compassionate, sympathetic, and kind thoughts in the mind towards this sick person.

The above version and its Pāli parallel do not proceed beyond depicting the mental attitude in response to seeing the sick traveler. This is unsurprising, as the episode is meant to illustrate the appropriate attitude toward someone thoroughly bad and thereby counter potential resentment; it is not meant to describe a course of action to be taken. Had the story been employed in a different context, it would quite probably have continued by showing the compassionate person doing whatever possible to ensure that the sick traveler indeed receives the type of assistance needed (Anālayo 2019c).

This suggestion finds confirmation in the description of the monastic who released an animal caught in a trap. These two cases are in fact fairly similar, insofar as both involve chancing across another who is suffering. The main difference is

---

[30] The description given here relates to a point made in the previous chapter (see above p. 71 note 47), in that the image of villages found in close proximity to each other along a road would make travel in the ancient Indian setting a safe undertaking, forestalling the kind of predicament described in the above passage.

that in the previous case the one who suffers is an animal rather than a human being.

In the above description, compassion finds expression in the witnessing person's wish for the sick person to be in some way relieved from misfortune. In other words, although the vision of the ailing traveler's distress forms the starting point, compassion takes as its main object the positive idea of the sick person being helped and finding relief. The task is not to keep dwelling on the actual suffering, imagining in detail how it must feel to be sick and helpless in this way. Instead of dwelling on the pain of the other, the cultivation of compassion takes as its object the potential alleviation of the pain.

Compassion without Grief

As the sincere aspiration for the absence of harm, compassion extends to all sentient beings and thereby naturally also includes oneself. Moreover, as just mentioned, the actual cultivation of compassion proceeds from the empathic apperception of suffering to the aspirational vision of the other being free from affliction. This puts into perspective a tendency found in some later Buddhist traditions to conceive of compassion as an act of taking upon oneself the pain and suffering of others (Anālayo 2017c: 99–103).

Research in cognitive psychology has shown that the cultivation of empathy, in the sense of taking upon oneself the pain of others, activates different brain areas and has divergent effects on the body compared to compassion (Dahl et al. 2015: 519). Focusing on the stress in another individual relates to elevated cortisol levels, but compassion rather leads to lower levels of cortisol reactivity (Buchanan et al. 2012). This supports the relevance of drawing a clear distinction between compassion as such and the experience of pain on behalf of others.

The need for such a clear distinction can be seen in the following statement by the Dalai Lama (2011: 55), who notes that "although compassion arises from empathy, the two are not the same. Empathy is characterized by a kind of emotional resonance—feeling *with* the other person. Compassion, in contrast, is ... wishing to see them relieved of their suffering ... compassionate doctors would not be very effective if they were always preoccupied with sharing their patients' pain. Compassion means wanting to *do* something to relieve the hardship of others."

Theravāda exegesis takes a comparable position, which finds its expression in the notion of "near enemies" to each of the four divine abodes. The idea is that certain unwholesome states can be mistaken for being the desired wholesome quality. In the case of compassion, the near enemy is worldly grief.[31]

In the face of the catastrophic repercussions of climate change and the vast scale of suffering, avoiding this near enemy is easier said than done. Here, it can be particularly helpful to rely on mindfulness. Just as mindfulness enables being with physical pain without either switching off or else resisting (Anālayo 2016d), similarly mindfulness can ease the mental pain of facing the horror of what human beings are doing to themselves and other sentient beings on this planet.[32] Relying on mindfulness as the main tool ensures that any grief or sadness that has manifested is not just being suppressed. Instead, it is witnessed but not acted out.

---

[31] Vism 319,13 (translated by Ñāṇamoli 1991: 311).

[32] Kaza 2018: 433 comments on "the Western mindfulness movement" that "to date, these trainings have been minimally applied to environmental contexts, but they may prove effective in future environmental disasters." Examples would be the benefits of mindfulness for professional firefighters or for children who have survived a hurricane, on which see Smith et al. 2019 and Cutright et al. 2019 respectively.

In terms of a cultivation of the divine abodes, mindfulness signals when it is time to shift from compassion to equanimity. For the cultivation of equanimity, some of the passages surveyed earlier can be relied on.

The Discourse on the Elephant's Footprint establishes a relationship between the earth element inside of one's body and the external earth, in order to drive home the truth that both are impermanent. Awareness of impermanence can serve as a powerful tool to cultivate equanimity, based on the realization that, whatever one experiences,[33]

> is of an impermanent nature, of a nature to cease, of a nature to decline, and of a nature to change.

Insight into impermanence can in turn lead to a growing appreciation of the empty nature of all phenomena. This finds expression in the same Discourse on the Elephant's Footprint in the indication that one who practices in this manner

> does not have this thought: "This is me; this is mine; I belong to it." How could one have such a thought?

In this way, based on being alerted to the need for equanimity through mindful monitoring, practice can be adjusted by relying on the above reflections in order to handle the arising of grief skillfully.

In addition to monitoring such shifting of mental attitudes, the cultivation of mindfulness itself has freedom from sadness as its aim. This comes up right away in the introductory part of the Discourse on the World Ruler, according to which mindful self-reliance involves the following:[34]

---

[33] MĀ 30 at T I 464c13, parallel to MN 28 at MN I 185,30.

[34] DĀ 6 at T I 39a28, parallel to DN 26 at DN III 58,13 and MĀ 70 at T I 524c15, which seems to give only an abbreviated reference to

removing greed and sadness in the world.

The same topic features also in a statement on the purpose of cultivating the establishments of mindfulness. The four establishments of mindfulness present a meditative path that has, among its aims, the following:[35]

for the purification of sentient beings, for causing the transcendence of worry and sorrow, for extinguishing affliction and pain.

In this way, the texts clearly see mindfulness as having the potential to enable transcending sorrow and grief.

With the help of mindfulness, it can at times be possible to transform grief into a sense of urgency (saṃvega). Giustarini (2012: 522) explains that "urgency ... is important in inspiring a conversion from indulgence in defilements to endeavour in the path, and this function is important not only at the beginning of the path or before, but whenever defilements prevail."[36]

Compassion in the Eightfold Path

Besides serving to counter the near enemy of compassion, mindfulness stands in a close relationship to the cultivation of compassion and mettā.

This can be seen through a closer look at the eightfold path. The second factor of this path, right intention, can take the form of the absence of ill will and of cruelty. These have their positive counterparts in mettā and compassion. From this viewpoint,

---

the four establishments of mindfulness and thus has no counterpart to the above quote.

[35] SĀ 607 at T II 171a10, parallel to SN 47.1 at SN V 141,10 (translated by Bodhi 2000: 1627).

[36] On saṃvega see also Coomaraswamy 1943.

mettā and compassion can be related in particular to the second factor of the eightfold path.

The eight factors of this path relate to each other in two complementary ways. The first of these two ways is sequential, in that each factor builds on the preceding ones:[37]

> Right view gives rise to right intention, right intention gives rise to right speech, right speech gives rise to right action, right action gives rise to right livelihood, right livelihood gives rise to right effort, right effort gives rise to right mindfulness, and right mindfulness gives rise to right concentration.

From the viewpoint of *mettā* and compassion, this would imply that their cultivation lays a foundation for the seventh path factor of right mindfulness.

The discourse that forms the source of the above quotation also presents a collaborative perspective, where right view, right effort, and right mindfulness serve to establish other path factors. Here right view is responsible for discerning whether a particular path factor is of the right or wrong type. Right effort steps in to sustain the former and eliminate the latter, while right mindfulness supervises. It is through a collaboration of these three path factors that a path factor like right intention can become established.[38] In this way a dynamic collaboration of the path factors emerges, where right view continuously sets the proper course and right mindfulness with equal continuity monitors what is

---

[37] MĀ 189 at T I 735c8, parallel to MN 117 at MN III 76,1 (translated by Ñāṇamoli 1995/2005: 938), and Up 6080 at D 4094 *nyu* 46b2 and P 5595 *thu* 86a6; for a comparative study of MN 117 see Anālayo 2010 and 2011b: 657–664.

[38] MĀ 189 at T I 736a6, parallel to MN 117 at MN III 73,21 and Up 6080 at D 4094 *nyu* 45a3 and P 5595 *thu* 84b5.

taking place. In short, mindfulness is relevant before, during, and after the cultivation of the divine abodes (Anālayo 2019c).

The Ethical Dimension of the Divine Abodes

In the context of the eightfold path, right intention reveals the relevance of *mettā* and compassion for ethical conduct. A complementary perspective on such relevance emerges also in the Discourse on Intention, which sets the divine abodes in opposition to the ten unwholesome courses of action. The discourse introduces these ten in this way:[39]

> Herein, there are three [types] of intentionally performed bodily deeds that are unwholesome, that result in the experience of *dukkha* and that have *dukkha* as their fruit; there are four [types] of verbal deeds and three [types] of mental deeds that are unwholesome, that result in the experience of *dukkha* and that have *dukkha* as their fruit.

By way of recapitulation, unwholesome bodily deeds comprise killing, stealing, and sexual misconduct; the four verbal deeds are false speech, malicious speech, harsh speech, and gossiping; and the three courses of action belonging to the mental realm comprise greedy desires, ill will, and wrong view.

---

[39] MĀ 15 at T I 437b28, parallel to AN 10.207 at AN V 297,20 (translated by Bodhi 2012: 1540) and Up 4081 at D 4094 *ju* 236b3 or P 5595 *tu* 270a6 (translated by Martini 2012; see also Dhammadinnā 2014). The later part of MĀ 15 instead parallels AN 10.208 at AN V 299,16 (translated by Bodhi 2012: 1542). The two Pāli parallels appear to have come into being due to an error during transmission, as a result of which AN 10.207 parallels the present exposition of the courses of action in MĀ 15 and Up 4081, whereas AN 10.208 parallels the exposition of the divine abodes in MĀ 15 and Up 4081, to be taken up below; for a discussion and translation of MĀ 15 see Anālayo 2009a.

After having expounded in detail these ten unwholesome courses of action, the Discourse on Intention continues like this:[40]

> A learned noble disciple leaves behind unwholesome bodily deeds and cultivates wholesome bodily deeds, leaves behind unwholesome verbal and mental deeds and cultivates wholesome verbal and mental deeds.
>
> Being endowed with diligence and virtue in this way, having accomplished purity of bodily deeds and accomplished purity of verbal and mental deeds, being free from ill will and free from contention, discarding sloth-and-torpor, being without restlessness or conceit, removing doubt and overcoming arrogance, [the learned noble disciple] is with right mindfulness and right knowing, without bewilderment.
>
> That [learned noble disciple] dwells accomplishing pervasion of one direction with a mind imbued with *mettā*, and in the same way the second, third, and fourth directions, the four intermediate [directions], above and below, completely and everywhere.
>
> Being without mental shackles, without resentment, without ill will, and without contention, with a mind imbued with *mettā* that is supremely vast and great, boundless and well cultivated, [the learned noble disciple] dwells accomplishing pervasion of the entire world.

The above description corresponds to the standard way in which the early discourses describe the meditative cultivation the divine abodes. I will return to the significance of this description in the next section of this chapter.

---

[40] MĀ 15 at T I 438a3, parallel to AN 10.208 at AN V 299,16 and Up 4081 at D 4094 *ju* 238a2 or P 5595 *tu* 271b8.

From the viewpoint of ethics, it is significant that this passage highlights the need for a firm foundation in moral conduct in order to be able to cultivate meditation successfully. This reflects a general pattern in early Buddhist thought, where the building of an ethical foundation occupies a central position.

At the same time, the relationship of ethics to the divine abodes is a reciprocal one and not confined to the former building the foundation for the latter.

This much can be seen from the ensuing exposition in the Discourse on Intention:[41]

> [The Buddha said]: "That [learned noble disciple] reflects like this: 'Formerly my mind was narrow and not well cultivated; now my mind has become boundless and well cultivated.'
>
> "When the mind of the learned noble disciple has in this way become boundless and well cultivated, if because of [associating with] bad friends one formerly dwelt in negligence and performed unwholesome deeds, those [deeds] cannot lead one along, cannot defile one, and will no longer follow one.
>
> "Suppose there is a small boy or girl who since birth has been able to dwell in the liberation of the mind through *mettā*. Later on, would they still perform unwholesome deeds by body, speech or mind?"
>
> The monastics answered: "Certainly not, Blessed One."
>
> [The Buddha said]: "Why is that? They do not perform bad deeds themselves, [so] how could [results of] bad deeds arise? Therefore, a man or a woman, at home

---

[41] MĀ 15 at T I 438a11, parallel to AN 10.208 at AN V 299,23 and Up 4081 at D 4094 *ju* 238a5 or P 5595 *tu* 272a5.

or gone forth,[42] should constantly make an effort to cultivate liberation of the mind through *mettā*."

This description, similarly applied to the remaining three divine abodes, reveals the other side of the relationship between
ethics and the divine abodes. The one side is the support moral
conduct provides for the cultivation of the divine abodes; the
other side is the impact of their cultivation on ethics in turn.

One who truly cultivates *mettā* and the other divine abodes
becomes increasingly unable to act in unwholesome ways. This
is the case to such an extent that, had one begun such cultivation from childhood onwards, one would no longer act in unwholesome ways. For this reason, everyone should constantly
make an effort to cultivate these beneficial mental states.

The Radiation of the Divine Abodes

The early discourses regularly describe the meditative cultivation of the divine abodes as a radiation in all directions, similar
to the passage translated above. The unlimited nature of this
radiation, free from any obstructions, finds reflection in their
qualification as "boundless" or "immeasurable" states.

Later exegetical tradition considers such radiation to be invariably descriptive of absorption attainment. This does not
seem to hold for the early discourses, where a radiation of the
divine abodes can take place even at levels well below absorption (Anālayo 2015b: 20–26).

One instructive episode concerns a brahmin who is taught,
apparently for the first time in his life, the meditative cultivation
of the divine abodes while on his deathbed. Previously, this brahmin had been engaging in immoral conduct. The way the dis-

---

[42] This distinction is not mentioned explicitly in AN 10.208; it is
found in Up 4081 at D 4094 *ju* 238a7 or P 5595 *tu* 272a7.

course describes him does not give the impression that he was a meditator, let alone someone with the meditative expertise required for the attainment of absorption. After being in grave pain until the time of his death, according to the discourse's report he was reborn in a celestial realm, the so-called Brahmā world.[43]

Such a rebirth fits the case of someone who has successfully cultivated the divine abodes, which implies that the brahmin had indeed been able to make good use of the instructions on their radiation he had just received. It follows that descriptions of such radiation appear to be indeed about how to cultivate the divine abodes and not only about full absorption attainment.

In later Buddhist traditions, however, the approach to meditation on the divine abodes tends to focus on individual persons as objects, and the radiation requires first developing this person-oriented approach (Anālayo 2015a). For the case of *mettā*, for example, Pāli exegesis recommends that meditative cultivation should start by taking first oneself as the object, then someone dear, then someone neutral, and finally a hostile person.[44]

A person-oriented approach to the cultivation of the divine abodes is also found in other Buddhist traditions, although at times recommended in particular for those unable to practice the radiation due to the presence of some defilement in the mind.[45]

In the early discourses, the meditative radiation of the divine abodes finds the following illustration:[46]

---

[43] MN 97 at MN II 195,22 (translated by Ñāṇamoli 1995/2005: 796) and its parallel MĀ 27 at T I 458b13; for a comparative study see Anālayo 2011b: 566–572.

[44] Vism 297,18 (translated by Ñāṇamoli 1991: 290).

[45] See the *Abhidharmakośabhāṣya*, Pradhan 1967: 454,7 (translated by Pruden 1988: 1269), and for a detailed discussion Dhammajoti 2010.

[46] MĀ 152 at T I 669c10, parallel to MN 99 at MN II 207,22 (translated by Ñāṇamoli 1995/2005: 816); see Anālayo 2011b: 572–579.

It is just as if, in an area where nobody has ever heard it, a person skilled at blowing a conch mounts a high hill at midnight to blow a conch with utmost effort and a wonderful sound comes out of it that pervades the four directions.

In the ancient Indian setting, blowing a conch was regularly used for religious ceremonies and also served as a means of communication over longer distances. It was even employed in warfare for such purposes, as its penetrative sound stands a good chance of being heard over the din of a battlefield. The blowing of a conch requires skill and is not achieved merely through effort. The lips need to be puckered appropriately to create the vibratory effect that will then result in the conch producing sound. Once this has been achieved, the resultant sound is both penetrative and beautiful. Presumably, the combination of pervasiveness in all directions and beauty made conch blowing an apt illustration for the radiation practice.

The boundlessness of the resultant experience of the divine abodes also aligns with mindfulness, which can lead to a similarly broadly open, and even boundless, state of mind. This becomes evident in some depictions of mindfulness of the body. Such descriptions appear to involve a rooting of mindfulness in the presence of the body while at the same time remaining openly receptive to whatever may happen at any of the bodily sense doors. Mindfulness practice undertaken in this way results in an "immeasurable" or "boundless" condition of the mind,[47] thereby being qualified with the same term that serves as a designation for the divine abodes.

The Experience of the Divine Abodes

In what follows, I intend to round off the exploration in this chapter by illustrating the distinct quality of each divine abode, on the understanding that this is based on my own ideas and

---

[47] MN 38 at MN I 270,11 (translated by Ñāṇamoli 1995/2005: 360) and its parallel MĀ 201 at T I 769c16.

not something found explicitly in the early discourses. My illustration is based on the example of the sun, on the assumption that one is in a cool and stable climate where sunshine is experienced as something pleasant and agreeable, instead of being oppressive.

Here, *mettā* would be like the sun at midday in a cloudless sky, equally illuminating everyone and providing warmth in all directions. Just as the sun shines on what is high and low, clean and dirty, so *mettā* shines on all without making distinctions.

The sun keeps shining, regardless of how its rays are received. It does not shine more if people move out into the open to be warmed by its rays, nor does it shine less because people move back indoors. Similarly, *mettā* does not depend on reciprocation. Its rays of kindness shine on others out of an inner strength that pervades all bodily activities, words, and thoughts, without expecting anything in return. From the center of one's being, *mettā* shines its rays on anyone encountered, just as from the midst of the sky the sun shines in all directions.

Compassion could be compared to the sun just before sunset. Darkness is close by, almost palpably close, yet the sun keeps shining. In fact, it shines all the more brilliantly, beautifully coloring the sky at sunset. Similarly, when face to face with suffering and affliction, the mental attitude of compassion shines even more brilliantly, undeterred by all the darkness found in the world.

Continuing with the sun imagery, sympathetic joy could be compared to early morning sunrise. The birds are singing merrily, the air is fresh, and the surroundings are illuminated by the rising sun and appear as if pervaded by joyful delight. At times the rays of the sun touch a dewdrop on a flower or tree and break into a myriad of colors. In the same way, the manifestation of wholesome qualities can become the source of joyful rejoicings within oneself.

The fourth of the divine abodes, equanimity, could then be likened to the full moon on a cloudless night. Just as the sun and the moon are both up in the vast sky, in the same way the four divine abodes share with each other the boundless nature of a mind that has become limitless like the sky. The moon is not itself a source of sunlight, unlike the sun. At the same time, however, the moon does reflect the light of the sun, just as equanimity reflects within itself the positive disposition of the other three divine abodes.

Summary

Early Buddhist thought recognizes the divine abodes as temporary states of liberation alongside the irreversible liberation attained with the stages of awakening.

The divine abodes, as one type of such temporary liberations, have a close relationship with ethical conduct. Their cultivation can in several ways support facing the challenges of climate change, which holds in particular for compassion as the wish for the absence of harm.

Throughout, mindfulness serves as a key quality, which in various ways relates to the cultivation of divine abodes like *mettā* and compassion. Practiced in conjunction, mindfulness and the divine abodes provide the fertile ground for growth in the type of inner balance and resiliency that the current environmental crisis requires.

Walking the Path

What is reckoned to be the truth of [the path] leading out of *dukkha*? That is, it is the noble eightfold path, namely right view, right thought, right speech, right action, right livelihood, right effort, right mindfulness, and right concentration.[1]

The teaching of the fourth truth needs to be put into practice. This takes the form of an eightfold path that has progress to awakening as its overarching aim. In this chapter, a natural focus in my exploration will be to apply the eightfold path to the challenges of climate change.

At the outset, however, I first attempt to relate the current environmental crisis to progress to awakening, based on a discourse that depicts a future time when the whole earth will be completely destroyed. In the final part of this chapter, I turn to mindfulness of death as a practical way to train oneself in the type of mental resilience required to face the catastrophic repercussions of climate change.[2]

Awakening

In order to do full justice to the overall soteriological orientation of early Buddhism and provide the appropriate setting for linking climate change to the different practices that make up the eightfold path, I need to relate the eightfold path's overall orientation toward awakening to environmental concerns.

---

[1] EĀ 24.5 at T II 619a17.

[2] On the employment of the term "resilience" in relation to climate change see Klepp and Chavez-Rodriguez 2018: 16–18.

Whereas in the previous chapters I tried to apply early Buddhist teachings to climate change, in what follows I try to do the reverse, in the sense of showing that mindfully facing the consequences of climate change can offer a substantial contribution to progress toward awakening. In this way, from having explored the environmental relevance of early Buddhism, I now turn to the soteriological relevance of facing climate change.

For the purpose of establishing this perspective, I take up the Discourse on Seven Suns, which describes a scenario much worse than anything that could result from climate change: a complete obliteration of the whole earth. In this discourse, the vision of such total destruction serves as a means to drive home the truth of impermanence and thereby lead onwards to awakening.

Drought

A Chinese version of the Discourse on Seven Suns begins by announcing the exposition's main theme of impermanence, followed by describing the onset of a drought in the following way:[3]

> All formations are impermanent, of a nature not to last, quickly changing by nature, unreliable by nature.[4] In this way, one should not delight in or attach to formations,

[3] MĀ 8 at T I 428c9, parallel to AN 7.62 at AN IV 100,5 (translated by Bodhi 2012: 1071, referred to as number 66), Sanskrit fragments, Dietz 2007 (perhaps not even worth additional mention is SHT VI 1267Bw, Bechert and Wille 1989: 57, identified in Bechert and Wille 2004: 414), T 30 at T I 811c24, EĀ 40.1 at T II 736b1, a complete discourse quotation in Up 3008, edited and translated by Dietz 2007, and a partial quotation Up 3094 at D 4094 *ju* 187b3 and P 5595 *tu* 214b1; see also Lamotte 1976: 2091f.

[4] The translation "unreliable" is based on adopting a variant reading.

one should [view] them as distressful, one should seek to
abandon them, one should seek to be liberated from them.
Why is that?

There will be a time when it will not rain. When it does
not rain, all the trees, the hundreds of grains, and all me-
dicinal shrubs will wither entirely, come to destruction
and extinction, unable to continue existing.

This is why [I say that] all formations are impermanent, of a nature not to last, quickly changing by nature,
unreliable by nature. In this way, one should not delight
in or attach to formations, one should [view] them as distressful, one should seek to abandon them, one should
seek to be liberated from them.

The impermanent nature of all formations is a recurrent theme
throughout the discourse, which repeats the last paragraph invariably after each of the different stages depicting how increasing degrees of heat impact the earth. The Pāli parallel proceeds
similarly but with a minor difference in formulation, as the lis-
tener is instead encouraged to become disenchanted, dispassionate,
and liberated. Here the reference to dispassion and disenchant-
ment would point to the same aim expressed in the above translated discourse in terms of not delighting in or attaching to formations. In both versions the final goal is to become liberated.

In this way, the description of the drought has a function
comparable to the depiction of the disappearance of the earth
in the Discourse on the Elephant's Footprint, taken up in the
first chapter of my study (see above p. 27). In that discourse,
the disappearance of the earth during a time of inundation similarly served to drive home the truth of impermanence.[5]

---

[5] On parts of the ensuing description that have similarities with the
present discourse see Anālayo 2011b: 197.

Desiccation

The Discourse on Seven Suns continues by describing how a progressive increase of heat impacts living conditions on the earth.[6] In the translation below, I have elided the repeated occurrences of the paragraph on impermanence, found after each depiction of increasing heat and its repercussions:

> Again, there will be a time when a second sun will appear in the world. When the second sun appears, the flow of all the streams and rivulets will become exhausted, unable to continue existing ...
>
> Again, there will be a time when a third sun appears in the world. When the third sun appears, all the great rivers will become exhausted, unable to continue existing ...
>
> Again, there will be a time when a fourth sun appears in the world. When the fourth sun appears, the great springs from which the five rivers of *Jambul* Island emerge ... those great springs will all become exhausted, unable to continue existing ...[7]
>
> Again, there will be a time when a fifth sun appears in the world. When the fifth sun appears, the water of the great ocean will recede ... and [eventually] there will be a time when the water of the ocean will be completely exhausted, being not sufficient to submerge [even] one finger.

As the title of the discourse indicates, its presentation involves altogether seven suns. In order to appreciate this description of

---

[6] MĀ 8 at T I 428c16, parallel to AN 7.62 at AN IV 100,20, Dietz 2007: 95,15, T 30 at T I 812a3, and EĀ 40.1 at T II 736b7.

[7] The *Jambul* fruit appears to be a black plum, *syzygium cumini*; see Wujastyk 2004.

the successive appearance of seven suns, it can again be helpful to take into account the symbolic function of numbers in
the setting in which the discourse was given. Dumont (1962:
73) notes that in its ancient Indian usage the number seven in
particular "also indicates a totality".[8] Rhys Davids and Stede
(1921/1993: 673) explain that, when appearing in Pāli texts,
seven "is a collective and concluding (serial) number; its application has spread from the *week* of 7 days" and its usage is
invested "with a peculiar magic nimbus".

A symbolic use of the number seven can be seen, for example, in the Discourse on the Establishments of Mindfulness.
The Pāli version and one of its two Chinese parallels depict
different time periods within which the systematic cultivation
of mindfulness can lead to complete freedom from greed and
aversion. Such freedom corresponds to reaching the higher two
levels of awakening, which result in becoming a non-returner
or an arahant.

The actual description of these time periods begins with
seven years and then counts down to six, five, four, three, two,
and one year, after which it continues with seven months, six,
five, four, three, two, and one month, followed by counting
down even further.[9]

The sequence of enumeration adopted here does not follow
numerical logic. Instead, it proceeds from seven down to one
in the case of both years and months. This is hardly meant to

---

[8] The symbolic function of the number seven in Buddhist texts has
also been noted by, e.g., Senart 1882: 285, Keith 1917: 408, Sen
1974: 64, and Gombrich 1975: 118.

[9] MN 10 at MN I 62,34 (translated by Ñāṇamoli 1995/2005: 155),
also in DN 22 at DN II 314,11 (translated by Walshe 1987: 350),
and their parallel MĀ 98 at T I 584b16. Another parallel, EĀ 12.1,
does not have such a listing of different time periods.

imply that it will be impossible to reach awakening after, say, eight months. Instead, the employment of seven serves to indicate a totality of time, and the ensuing countdown clarifies that such a complete time period, be it years or months, is not indispensable, as the same realizations can also take place earlier.

Applying this sense to the present context, the depiction of the arising of up to seven suns could perhaps be interpreted in a symbolic manner. On this assumption, the idea of an increase of heat until the earth becomes "totally" heated up might have found expression by depicting a series of seven suns rather than an increase in the size of a single sun. The shift from one to two suns could then be interpreted as suggestive of a doubling of the subjective experience of heat. In other words, the difference between the average bodily temperature and outside temperature during summer in India has become doubled.

On this interpretation, the cumulative effect of the arising of ever more suns then eventually leads to a stage at which the heat has become seven times as much as what would have been experienced as normal in the ancient setting.

Whether or not this suggestion captures the original intention, as a result of the increase of heat eventually the five great rivers of India, here referred to as "*Jambul* Island", dry up and all the water in the ocean disappears.

Conflagration

Despite the ocean water having all dried up, the increase of heat depicted in the Discourse on Seven Suns has still not come to its final climax:[10]

Again, there will be a time when a sixth sun appears in

---

[10] MĀ 8 at T I 429a12, parallel to AN 7.62 at AN IV 103,1, Dietz 2007: 96,36, T 30 at T I 812b13, and EĀ 40.1 at T II 736b22.

the world. When the sixth sun appears, the whole great earth and Meru, the king of mountains, will completely emit smoke, a single mass of smoke. It is just like a potter's kiln which, at the time of being ignited, will completely emit smoke, a single mass of smoke ...

When a seventh sun appears, the whole great earth, and Meru, the king of mountains, will be thoroughly burning, completely ablaze, a single mass of flame ...

When the seventh sun appears, Meru, king of mountains, together with this great earth, will burn down and be destroyed, with not even ash remaining.[11] It is just like burning ghee that is fried until being completely exhausted and not even smoke or soot remains ...

For this reason, [I said that] all formations are impermanent, of a nature not to last, quickly changing by nature, unreliable by nature. In this way, one should not delight in or attach to formations, one should [view] them as distressful, one should seek to abandon them, one should seek to be liberated from them.

I have now told you that Meru, king of mountains, will collapse and be destroyed. Who is able to believe this? Only those who have seen the truth.

I have now told you that the water of the great ocean will be exhausted and vanish. Who is able believe this? Only those who have seen the truth.

I have now told you that the whole great earth will be burnt up and extinguished. Who is able believe this? Only those who have seen the truth.

The description is based on ancient Indian cosmology, which places Mount Meru at the center of the world, surrounded by

---

[11] The translation "ash" is based on adopting a variant reading.

four continents, which are in turn surrounded by the ocean.[12] In the above description, the reference to Mount Meru as the "king of mountains", in the sense of being the largest among them and forming the central pillar of the ancient Indian cosmos, conveys the sense of the totality of the destruction. Even such an incredibly large mountain, the pole of the whole world in an ontological and cosmological sense, will be completely obliterated.

The gradual destruction of the earth due to increasing solar heat has similarities with current scientific knowledge.[13] According to Schröder and Smith (2008: 156 and 159), in the far distant future, "the time will come when the increasing solar flux will raise the mean temperature of the Earth to a level ... at which life is no longer sustainable."

Eventually, "the oceans will start to evaporate ... until the oceans have boiled dry ... the subsequent dry greenhouse phase will raise the surface temperature ... and the ultimate fate of the Earth, if it survived at all as a separate body ... would be to become a molten remnant."

The point of noting this similarity is not to pretend that early Buddhist thought already anticipated current scientific knowledge. Other cosmological elements reflected in the above discourse and elsewhere in the early texts prevent drawing such a conclusion.

Nevertheless, the fact that this description concords with what we currently would expect to happen to the earth in the distant future helps to make the imagery presented in the Dis-

---

[12] See in more detail, e.g., Kloetzli 1983: 23–45 or Sadakata 1997/ 2004: 25–40.

[13] The idea as such is fairly well-known in Buddhist circles; for example, Chokyi Nyima 2009: 138 comments that "the future world will become very hot—seven times the heat of the present sun."

course on Seven Suns come alive, together with its chief sote-riological message: all things are impermanent, so one should remove one's attachments, cultivate dispassion, and increas-ingly let go.

The early stages of the description in the Discourse on Seven Suns could be related more closely to the current situation. Whereas the eventual burning up of the whole earth is expected to take place in the distant future, a more imminent threat is the potential repercussions of methane release on global warming.[14]

Large amounts of methane are stored in arctic permafrost areas and now appear to be escaping at an increasing rate due to global warming. The eventual consequences of such a release might turn out to be similar to the greatest known mass extinction of species in the history of this planet, which appears to have occurred about 250 million years ago, at the end of the Permian period. Brand et al. (2012: 121) report that this event resulted in the extinction of approximately 90% of marine species and 70% of terrestrial vertebrate species.

According to an explanation advanced for this mass extinction, increased volcanic eruptions at that time led to an increase of carbon dioxide, resulting in global warming. The ensuing melting of permafrost led to the release of large quantities of methane, which accelerated global warming to such a degree that it resulted in the obliteration of the majority of species. Brand et al. (2016: 506) propose that "the emission of carbon dioxide from volcanic deposits may have started the world onto the road of mass extinction, but it was the release of methane from shelf sediments and permafrost hydrates that was the ultimate cause for the catastrophic biotic event at the end Permian [period]."

---

[14] See, e.g., Anthony et al. 2018, Knoblauch et al. 2018, and Sha-khova et al. 2019.

Applied to the current situation and expressed in terms of the Discourse on Seven Suns, although the arising of the seventh sun would only take place in the distant future, the arising of the second sun could happen relatively soon. From the viewpoint of sustainability of human life on earth, the stages leading up to the arising of the second sun are in fact decisive, as that much suffices for effecting the extinction of human life on earth.

Loy (2018: 135) reasons that, "if humanity disappears, Buddhism and all other religious traditions will also disappear, along with the compositions of the greatest musicians, the paintings and sculpture of the greatest artists, the poetry and plays and novels of the greatest writers, the discoveries of the greatest scientists, and so on. Take a moment to reflect on that ... how should we live today, in the face of that very real possibility?"

Facing the extinction of humanity and even of the whole earth relates to another noteworthy aspect of the above-translated version of the Discourse on Seven Suns, which is the mention of those who have seen the truth. According to the Pāli commentary, the corresponding expression in the Pāli discourse refers to those who have reached at least the first level of awakening, stream-entry.[15]

The reference to those who have seen the truth conveys the impression that the presentation in this discourse is meant to be taken literally. The Discourse on Seven Suns therefore appears to be similar in kind to the Discourse on the Elephant's

---

[15] The Pāli commentary on the corresponding expression *diṭṭhapada* in AN 7.62 at AN IV 103,23, Mp IV 52,9, explains that this stands for those who have reached stream-entry. AN 7.62 applies this statement only to the destruction of Mount Meru, not to the drying up of the great ocean or the extinction of the whole earth. The corresponding statement in the Tibetan version, Dietz 2007: 97,36, refers to believing in the burning up of Mount Meru and the great earth. T 30 at T I 812c3 has a reference to the destruction of the great earth and the heavens, although the preceding part lacks a reference to those who believe in this. EĀ 40.1 at T II 736c22 proceeds differently from the other versions. After completing its description of the repercussions of the seventh sun, it continues by depicting a devastation caused by water and then by wind. This could be the result of a shifting of textual portions within this collection, a pattern discussed in Anālayo 2014/2015 and 2015c.

Footprint, in the sense of reporting something actually expected to happen rather than being a parable.

From the viewpoint of the Discourse on Seven Suns, only those who have reached at least stream-entry are able to take its message fully to heart. In an attempt at interpretation, perhaps having had a first experience of the cessation of the dependent arising of *dukkha* at stream-entry is what enables placing trust in the idea of the complete cessation of all dependently arisen life on earth.

Be that as it may, the vision of a total cessation of human civilization can serve as a training to inculcate the type of letting go that can lead to the breakthrough to the different levels of awakening.

The same can in turn be applied to climate change, whose dire outcomes fall short of the total destruction envisaged in the Discourse on Seven Suns. In other words, learning to face with mindfulness the repercussions to be expected on earth due to climate change has a considerable potential in furthering progress to awakening. This potential relies on strengthening insight into impermanence and thereby bringing about dispassion and a thorough letting go.

Future Decline

The cultivation of insight into impermanence in this way is quite different from a fatalist attitude,[16] just as equanimity dif-

---

[16] *Pace* Keown 2007: 97, who reasons that "in Buddhism there seems to be an acceptance that, even an expectation that, the world will decline. This is thought of as a basic characteristic of the cosmic order: the eventual destruction of the environment is a basic feature of *saṃsāra*, and is exactly what we should expect. Efforts to prevent it may therefore be seen as naïve and deluded and contrary to a proper understanding of Dharma, or natural law. Against this

fers from indifference. This can be explored in relation to another and related topic that occurs with more frequency in the discourses than the destruction of the earth, namely the eventual vanishing of the Buddha's teachings.

Nattier (1991: 286) summarizes the message, in the form in which it emerges in one particular instance of such a prediction, as follows: "the Buddhist religious tradition is no less transitory than other worldly phenomena ... while human efforts can contribute to its preservation or can hasten its demise, they cannot provide it with an eternality."

At the same time, however, passages in the early discourses concerned with such disappearance regularly offer indications on how this dire prospect can be prevented.[17] In other words, the point is decidedly not to encourage a fatalist attitude.

The same holds equally in regard to human life. Recollection of death by way of learning to accept one's own mortality, a topic to which I return at the end of this chapter, is one of the meditation practices recommended in the early discourses. Nevertheless, even those accomplished in this practice will

background I see no obvious basis on which to address specific ecological questions."

[17] SN 16.13 at SN II 225,8 (translated by Bodhi 2000: 681) and its parallels SĀ 906 at T II 226c15 and SĀ² 121 at T II 419c6, SN 47.22 at SN V 172,21 (translated by Bodhi 2000: 1651), SN 47.23 at SN V 173,14 (translated by Bodhi 2000: 1651) and its parallel SĀ 629 at T II 175b27 (which just speaks of decline in general, not specifically related to the Buddha's teachings), AN 5.201 at AN III 247,20 (translated by Bodhi 2012: 818), AN 6.40 at AN III 340,13 (translated by Bodhi 2012: 904), and AN 7.56 at AN IV 84,22 (translated by Bodhi 2012: 1059, referred to as number 59). An exception to this pattern, related to the founding of the Buddhist order of nuns, can with considerable probability be set aside as a later development (Anālayo 2016b: 147–177).

continue to eat and attend to the needs of their bodies. As noted by Giustarini (2018: 1230), "the relevance of health, seen not just as a secondary form of wellbeing, but also as considerably functional to the cultivation of the factors of awakening, is often and variously emphasized" in the early texts.

Similarly, the vision of the complete destruction of the earth at some point in the future is not meant to lead to apathy. Instead, its function is to deconstruct tacitly held and unreasonable assumptions in order to introduce a realistic appraisal of the situation. This assessment then serves as a foundation for formulating an appropriate response to whatever situation presents itself now, before the complete destruction has transpired.[18]

Just as it is meaningful to look after the health of the mortal body in order to be able to practice, so it is meaningful to look after the living conditions on earth for the same purpose. The point is only that such endeavors should come with an inner balance due to the absence of attachment, in the knowledge that eventually the body will fall apart, the Buddha's teachings will disappear, and the earth will be completely destroyed.

Dispassion

Comparable to insight into impermanence, which is not meant to result in fatalism, so arousing dispassion, a key element of the early Buddhist path to awakening (Anālayo 2009c), does

---

[18] This holds not only from a Buddhist perspective. According to Moo 2015: 944f, "in the majority if not all of extant Jewish and Christian apocalypses in which future eschatology is a prominent feature ... the whole point is to give value and significance to how life is lived now." In fact, "a transformation of the imagination through apocalyptic narrative has the potential to motivate a commitment to creative and alternative ways of being in the world that prosaic arguments and mere consideration of scientific data does not."

not imply a stepping out of environmental responsibility. As explained by Sucitto (2019: 260), "because of that dispassion, one feels more agile, ready to participate without self-consciousness or ambition, more capable of cooperation."

Growing dispassion features as a central aspect of how the factors of awakening, a set of mental qualities whose chief purpose is to lead to awakening, should be developed. For this purpose, mindfulness as the first awakening factor should be cultivated in the following manner:[19]

> One cultivates the awakening factor of mindfulness supported by seclusion, supported by dispassion, and supported by cessation, conducing to letting go.

The same applies to the other awakening factors, each of which requires the support of seclusion, dispassion, and cessation in order to lead to letting go. A similar pattern of insight-related themes occurs as the concluding set of instructions for mindfulness of breathing. In fact, the passage quoted above also relates to mindfulness of breathing, here in particular showing how this practice can be combined with the arousing of the awakening factors in order to issue in liberation.

Different versions of the final set of instructions on mindfulness of breathing agree in setting out from the theme of impermanence, which is also the starting point in the Discourse on the Seven Suns. For the remaining instructions, the parallels present two different modalities. One of these proceeds from impermanence to eradication, dispassion, and cessation (Anālayo 2013b: 228–233). A sequence that corresponds more closely to the pattern for cultivating the awakening factors can

---

[19] SĀ 746 at T II 198a8, parallel to SN 46.66 at SN V 132,2 (translated by Bodhi 2000: 1620).

be found in a text on monastic discipline, the Sarvāstivāda *Vi-naya*, where the instructions take the following form:[20]

> one should be mindful of breathing in and out single-mindedly, contemplating impermanence [by] contemplating change,[21] contemplating dispassion, contemplating cessation, and contemplating letting go.

Pāli discourses adopt the same sequence. The pattern that emerges in this way could be employed to implement the main teaching given in the Discourse on Seven Suns. The breath is indeed a convenient tool for this purpose, as it exemplifies the dependency of the human body on external living conditions.

As mentioned in the first chapter, to stay alive the body requires a continuous supply of oxygen through the process of breathing. This will no longer be available with the gradual destruction of the earth due to increasing solar heat, described in the Discourse on the Seven Suns. Hence the cultivation of impermanence, dispassion, cessation, and letting go, in the way described in the above passage for mindfulness of breathing, could be employed to put into practice the insight-related import of the Discourse on Seven Suns and thereby actualize its awakening potential.

View

For the cultivation of such meditation practice, the factors of the eightfold path present the required context. This involves a

---

[20] T 1435 at T XXIII 8b1, parallel to Vin III 71,10 (translated by Horner 1938/1982: 122); for a comparative study of the preceding narration in a range of different *Vinaya*s see Anālayo 2014b.

[21] The double reference in the original to "impermanence" and "change" is probably the result of a textual error; see Anālayo 2016d: 249 note 16.

collaboration between establishing the proper perspective; undertaking communications, activities, and livelihood informed by that perspective; and engaging in meditative cultivation of the mind.

As already mentioned earlier, the early discourses describe the first path factor of right view in two different manners (see above p. 66). One of these definitions can be understood to offer the significant indication that one needs to take responsibility for one's actions, which in view of human-caused environmental destruction acquires additional significance.

The other definition requires adopting the perspective of the four noble truths. This alternative presentation implies that right view, as the first factor of the path described under the heading of the fourth truth, corresponds to all four noble truths.

Far from involving a tautology, the point appears to be that some basic appreciation of the four truths is required to generate motivation to set out on the path. Such a preliminary understanding of the four truths as a guiding principle for the practice of the path differs from the level of insight into the four truths gained at stream-entry or with full awakening. As the first factor of the eightfold path, right view in the form of the four truths can serve a preliminary diagnostic function. It places the disconcerting recognition of the fact of *dukkha* within a framework that reveals its conditionality as well as the possibility of becoming free from it, together with the means required to achieve that goal.

According to the Discourse on Turning the Wheel of Dharma, each of the four truths calls for a different activity. First of all, *dukkha* should be "penetratively understood", which in the approach presented here would be in particular understanding the dependency of the human body on conducive living conditions on the earth.

The arising of *dukkha* should be "abandoned". In the presence context, this can be taken to highlight the need to abandon the three root defilements, whether they manifest internally or externally.

The cessation of *dukkha* should be "realized". The Discourse on Seven Suns shows that progress to the realization of awakening is possible, based on cultivating a vision of the ultimate disappearance of the whole earth. By way of contribution to the eventual realization of full awakening, a direct contrast to the three root defilements can be found in the cultivation of the divine abodes as temporary liberations whose 'realization' is possible well before full awakening has been reached.

The path should be "cultivated". The details of such cultivation will emerge during the next pages. Suffice it for now to note the main point: the need to put it all into practice.

Intention

Right intention can take the following form:[22]

> Thoughts of dispassion, thoughts of non-ill will, thoughts of non-harming: this is reckoned right intention.

The Pāli formulation of the first of these three modalities of right intention speaks more specifically of "renunciation". This wording seems particularly appropriate to the challenge of climate change, the mitigation of which does require the willingness to renounce certain types of comfort in order to reduce one's carbon footprint. Sucitto (2019: 64) clarifies the

---

[22] MĀ 189 at T I 736a1, parallel to MN 117 at MN III 73,4 (translated by Ñāṇamoli 1995/2005: 935) and Up 6080 at D 4094 *nyu* 45a1 and P 5595 *thu* 84b2, which has a somewhat different listing, perhaps the result of a textual error; see the translation in Anālayo 2019g: 1932.

basic principle involved, in that "renunciation is the process of separating wants from needs."

The other two modalities of right intention concern the absence of harm and the absence of ill-will, which can provide a guideline for any type of activism. The intention to avoid harming could become a strong motivation to take fully into account the consequences of one's environmental habits on other humans and animals, in particular those less able to insulate themselves from the repercussions of global warming.

At the same time, it is also important that a response to the current crisis, however pressing it appears, stays within the framework set by non-harm and non-ill will. These two types of intention are in turn closely related to the cultivation of *mettā* and compassion, discussed in the previous chapter. In fact, one who has fully cultivated these divine abodes will no longer be dominated by the wish to harm or by ill will, or even righteous anger. When viewed from this perspective, the meditative practice of *mettā* and compassion can make a rather substantial contribution in support of right intention, which in turn provides the orientation for the ensuing path factors.

Speech

Right speech delineates four types of conversation better avoided:[23]

> Abstention from false speech, from divisive speech, from harsh speech, and from frivolous speech:[24] this is reckoned right speech.

---

[23] MĀ 189 at T I 736a10, parallel to MN 117 at MN III 73,30 and Up 6080 at D 4094 *nyu* 45a5 and P 5595 *thu* 84b7.

[24] The translation "frivolous" follows a correction in the CBETA edition, in line with the formulation employed in the preceding line for the corresponding wrong speech.

The canonical definition of right speech, which corresponds to the four courses of verbal action discussed earlier (see above p. 65), sets the parameters for any communication. Effectively communicating the complex topic of climate change, in particular, can be quite a challenge (Filho et al. 2019). From an early Buddhist perspective, first and foremost one should stick to the truth. This can take the form of countering misleading statistics, falsification of data, and corporate-sponsored research to spread falsehood without succumbing to intentional exaggeration and reliance on unsubstantiated information when trying to inform others.

Another aspect of relevance to ecological destruction would be to express the truth of one's inner conviction. Of considerable importance here could be participation in elections, which relates to the next path factor of right action, making sure that the opportunity to vote for whoever best represents one's opinion on the need to protect the environment is not missed out. In the USA, for example, nearly half of those in principle eligible to participate in elections do not vote. No wonder when eventually governments come into power that do not adequately reflect the opinion of the majority of the people.

In addition to voting, there are a range of other opportunities can be used to draw attention to the truth of climate change and the need to take action.

Moving on to the next aspect of right speech, when drawing attention to climate change, communications that create divisions are better avoided. In the original context, the import appears to be some form of backbiting that sets people against each other. But in a more general sense it could perhaps also be taken as discouraging the dualistic creation of clear-cut factions, such as contrasting those bad ones, held entirely responsible for the climate crisis, to the good ones who are opposing ecological destruction. This is not to say that differences do not exist, but only to suggest that communication can be undertaken in such a way as to avoid excessive emphasis on dualistic contrasts where the communicator is, of course, always on the good side.

Harsh speech is speech that hurts and therefore is directly opposed to the right intention for non-harm. Even face to face with those most directly responsible for climate change denial and destruction of the environment, it should still be possible to communicate from a point of balance that desires the absence of harm for everyone.

In the original setting, frivolous speech stands for conversations about irrelevant matters not related to the path of practice. Applied to the case of climate change, it could perhaps be taken as an encouragement to make the present ecological challenge a recurrent topic of conversation, to communicate relevant information and, perhaps even more importantly, to communicate a balanced attitude in facing the crisis. This could also entail aiming communications at producing concrete results, which in a debating situation would require listening and understanding the opposing side, taking their concerns into account, rather than just wanting to demolish the opponent's argument.

A pervasive problem, as noted by Filho (2019b: 2), is that "the substantial amount of information and knowledge on aspects of climate change mitigation and adaptation is not matched by a similar body of expertise on climate change communication, despite the fact that the latter is central to any attempts to modify the human behaviours and man-led actions which have been largely responsible for the climate-related problems seen today."

Of interest here could be a finding by Ranney and Clark (2016) that the majority of people have no adequate understanding of the basic mechanism involved in global warming.

Based on receiving such information, a change in attitude occurred toward more acceptance of the fact of global warming as well as willingness to act in ways that will mitigate it.

The basic information required for this purpose was just the following (Ranney and Clark 2016: 51f): "The earth transforms sunlight's visible light energy into infrared light energy, which leaves earth slowly because it is absorbed by greenhouse gases. When people produce greenhouse gases, energy leaves Earth even more slowly—raising Earth's temperature."[25]

When spreading information about the current crisis, it is particularly important to beware of "catastrophe fatigue". As noted by Tokar (2018: 182), "while some authors focus on the most dire future scenarios, hoping that people can be shocked into realizing the magnitude of changes that are necessary, this approach appears more likely to inspire despair and withdrawal than meaningful action." In the words of Hoggett (2011: 261), "the quandary we face is how to sound the alarm without being alarmist."

The problem is that "fear-related messages can lead to counter-productive behaviors," as observed by Troillet et al. (2019: 56), as "fear is an emotion that induces stress-related responses within the organism, leading to automatic reactions such as fight or flight or freeze. Those reactions may be adaptive in some situations such as fighting with a dangerous animal, but won't be effective in new situations requiring calm and positive emotions for creative responses."

---

[25] See also Filho 2019a for a survey of helpful strategies developed for the International Climate Change Information Programme, "the world's largest non-government funded information, communication and education programme on climate change" (p. 10), and Otto et al. 2019 on the potential and challenges of online courses to promote a better understanding of climate change and its repercussions.

Yuen (2012: 20) explains that "apocalyptic imagery has saturated popular culture for decades … the ubiquity of apocalypse in recent decades has led to a banalization of the concept—it is seen as normal, expected, in a sense comfortable," hence "environmentalists and scientists must compete in this marketplace of catastrophe, and find themselves struggling to be heard above the din."

Moreover, "right-wing and nationalist environmental politics have much more to gain from an embrace of catastrophism. This is especially true if the invocation of fear is the primary rhetorical device … [as] fear is not a stable place to organize a radical politics, but it can be a very effective platform from which to launch a campaign of populist xenophobia or authoritarian technocracy under the sign of scarcity."

Yuen (2012: 16) even warns that "environmental catastrophism may end up exacerbating the very problems to which it seeks to call attention."

At the same time, however, Moo (2015: 938) argues that "it would be a mistake to dismiss the potential of the apocalyptic imagination to offer ways of re-envisioning and re-engaging with climate change and the questions of meaning and value that it raises for us," serving "not merely as stereotypical fantasies of the 'end' or guilt-inducing narratives of inevitable disaster but rather as dramas that invite us to consider afresh who we are, where we are, and what we value."

Such reconsideration is indeed the purpose of the teaching provided in the Discourse on Seven Suns. What emerges in this way is the need to be circumspect in the deployment of catastrophe rhetoric and, perhaps even more importantly, to provide tools that enable the audience to actualize the positive potential of disaster imagery.

Action

The definition of right action corresponds to the first three of the five precepts usually taken by lay followers of the Buddha and to the three courses of bodily action discussed earlier (see above p. 65):[26]

---

[26] MĀ 189 at T I 736a18, parallel to MN 117 at MN III 74,24 and Up 6080 at D 4094 *nyu* 45b2 and P 5595 *thu* 85a5.

Abstention from killing, from taking what is not given, from sexual misconduct: this is reckoned right action.

Another passage shows a broad scope of application for such abstention, at the same time also providing a rationale for its undertaking. For the case of killing, this passage proceeds as follows:[27]

"If someone wishes to kill me, that is not enjoyable to me. What is not enjoyable to me is also like that for another. [So], how could I kill another?" Having had this reflection, one undertakes not to kill sentient beings and does not enjoy when sentient beings are killed.

The Pāli version explains that such a reflection motivates one to abstain from killing, to encourage others to abstain in the same way, and to speak in praise of such abstention. Both versions apply the same basic principle to the other two aspects of right action, abstention from theft and from sexual misconduct.

This sets a convenient frame for environmental activism, in the various forms in which this can manifest (Cassegård et al. 2017). It provides a clear-cut directive to act in the way one would wish others would act toward oneself. The basic rationale motivating such abstention can be applied in particular to the problem of climate justice, where the repercussions of climate change afflict most severely developing countries, who have contributed the least to causing them (Jacobson 2018 and Jafry 2019). It encourages placing oneself mentally into the situation of others and taking action based on that.

---

[27] SĀ 1044 at T II 273b16, parallel to SN 55.7 at SN V 353,29 (translated by Bodhi 2000: 1797); for a Gāndhārī parallel see Allon in Glass 2007: 12.

The passage also clarifies that abstention from killing, for example, can extend from not intentionally terminating the life of a sentient being oneself to avoiding even indirectly supporting the taking of life. An obvious application of that would be vegetarianism.[28]

The Karmapa (2013: 97f) comments that "our reliance on meat is a major cause of climate change, deforestation, and pollution ... the methane gases emitted by livestock contribute more to climate change than does carbon dioxide. This tells us that if we human beings made a significant shift toward becoming vegetarian, by that shift alone we could dramatically reduce global warming ...

"Vast quantities of feed, water, land, fuel, and other resources are required to sustain livestock ... studies indicate that the land needed to produce food for one meat-eater could support twenty vegetarians. This demonstrates how much smaller our ecological footprint could be just by giving up meat."

In addition to the harm caused to the environment in general and of course to the animals themselves, who are often raised in ghastly conditions and then cruelly slaughtered, eating meat also harms the consumer. Kaza (2008: 7) notes that "some vegetarians have turned away from meat to protect their health and avoid meat-associated medical risks. Studies now show that hormones used in beef production can affect human reproductive development, causing early puberty and male infertility. The heavy use of antibiotics in conventional meat and dairy operations is a human health concern as well, undercutting the effectiveness of these valuable drugs in treating human

---

[28] For a discussion of "meat-eating denial", in the sense of intentional spreading of misinformation regarding its environmental repercussions by those who would be affected if meat consumption were to be reduced, see Stanescu 2020.

infection. Reducing harm to ourselves is a viable and important aspect of reducing environmental impact."

For implementing such reduction of harm, a recommendation would be to start by adopting a meat-free diet one day of the week, for instance, and from that gradually increase the shift away from meat-consumption. Eventually it might also become possible to reduce reliance on other dairy products.

Livelihood

The articulation of right livelihood takes as its point of departure the situation of a monastic, evident in the following formulation:[29]

> If there is no seeking [requisites] with a dissatisfied mind, no having recourse to various inappropriate types of spells, no making a living by wrong forms of livelihood; if one seeks robes and blankets with what is in accordance with the Dharma, by means of the Dharma, seeks beverages and food, beds and couches, medicine and any [other] requisites of life with what is in accordance with the Dharma, by means of the Dharma: this is reckoned right livelihood.

The Pāli parallel to this passage is less detailed but conveys a similar sense. Proceeding from the livelihood of a mendicant monastic to that of a lay person, another Pāli discourse can be consulted, of which no parallel appears to be extant. The discourse lists five trades that a lay disciple of the Buddha should not engage in. These are trading in weapons, in sentient beings, in meat, in intoxicants, and in poisons.[30] The commentary adds

---

[29] MĀ 189 at T I 736a29, parallel to MN 117 at MN III 75,15 and Up 6080 at D 4094 *nyu* 46a2 and P 5595 *thu* 85b5.

[30] AN 5.177 at AN III 208,15 (translated by Bodhi 2012: 790).

that this refers not only to undertaking such trade oneself but also to getting others to undertake it.[31]

A central principle behind most occupations mentioned in this list appears to be the infliction of harm. Based on the conception of livelihood that emerges in this way, an application to the current crisis could revolve around minimizing harm to other sentient beings and the environment, be this caused directly by one's actions or indirectly. On this interpretation, any step taken in this direction could be considered an implementation of this particular dimension of the eightfold path.

Kaza (2018: 446f) reasons that, "for example, commuting to work by walking, biking, or mass transit would be a more virtuous choice than commuting by car or plane because one would personally be contributing less carbon pollution (and therefore less harm) to the atmosphere. Practising mindfulness in relation to energy use or carbon intensive products can be a path to individual liberation. One can study one's personal hooks or triggers related to maintaining privileged climate-controlled homes or one's desires for high carbon footprint food choices such as red meat and imported fruits ... this sort of personal practice is part of understanding the complexities of individual actions and belief systems that maintain currently unsustainable levels of carbon and other greenhouse gas emissions."

Effort

Right effort covers altogether four dimensions:[32]

---

[31] Mp III 303,21.

[32] SĀ 877 at T II 221a24, parallel to AN 4.13 at AN II 15,14 (translated by Bodhi 2012: 401), and Sanskrit fragments, SHT V 1445+ 1447, Sander and Waldschmidt 1985: 258f, and Hosoda 2003. A comparable description can be found in MĀ 189 at T I 736b9, which in this case is absent from MN 117; see Anālayo 2011b: 661.

One arouses desire, application, endeavor, and takes hold of the mind for abandoning already-arisen bad and un-wholesome states ... one arouses desire, application, endeavor, and takes hold [of the mind] for the non-arising of not-yet-arisen bad and unwholesome states ... one arouses desire, application, endeavor, and takes hold [of the mind] for bringing about the arising of not-yet-arisen wholesome states ... and one arouses desire, application, endeavor, and takes hold [of the mind] for the increase and cultivation of already-arisen wholesome states.

A significant indication offered by this description is that effort should first of all be directed inwards. This holds even for the current crisis whose speed and magnitude certainly call for quick action. Yet, from the viewpoint of the framework provided by the eightfold path, one needs to ensure that the condition of the mind is free from "bad and unwholesome states". Only once this is at least temporarily achieved has the time come to act on the external level.

Needless to say, the practice of right effort in this way continues during any activity undertaken on the external level. Based on envisioning the entire situation in terms of the three root defilements, as suggested above (see p. 77), such activity has as its central orientation the countering of their impact. Guided by this orientation, one's own internal condition will naturally be kept in view, so as to prevent the rearising of the three root defilements within.

Ñāṇaponika (1990: 8) reasons that, "if we leave unresolved the actual or potential sources of social evil within ourselves, our external social activity will be either futile or markedly in-complete. Therefore, if we are moved by a spirit of social re-sponsibility, we must not shirk the hard task of moral and spiritual self-development. Preoccupation with social activities must not be made an excuse or escape from the first duty, to tidy up one's own house first."

Mindfulness

The cultivation of mindfulness as an integral dimension of the path to awakening takes the form of four establishments:[33]

> There are four establishments of mindfulness. What are the four? They are reckoned to be the establishment of mindfulness by contemplating the body [in regard to] the body ... feeling tones ... the mind ... and the establishment of mindfulness by contemplating dharmas [in regard to] dharmas.

Such cultivation of the four establishments of mindfulness can become a regular meditation practice, in order to build up the mental resilience required for confronting the crisis.

Based on such formal cultivation, mindfulness can unfold its potential throughout any activity. This takes place by way of monitoring, in line with its role in relation to any of the other path factors, as discussed above (p. 109).

In this way, mindfulness can become a central tool for facing the horror of climate catastrophe with inner balance and, based on that, taking the steps needed to transform what might well be the most serious challenge human beings have ever faced in their history. With mindfulness, this challenge could

---

[33] SĀ 605 at T II 170c28, parallel to SN 47.24 at SN V 173,26 (translated by Bodhi 2000: 1652) and Up 6027 at D 4094 *nyu* 12b3 or P 5595 *thu* 45b6 (translated by Dhammadinnā 2018a: 23). These are listed under the heading of right mindfulness in MĀ 189 at T I 736b14, in which case again, as in the previous footnote, the parallel MN 117 does not provide such a definition.

be transformed into an opportunity, an opportunity to increase global awareness and move to a level of interaction among human beings that values the common welfare over individual profit in order to maintain the living conditions required for the survival of human civilization.

From the viewpoint of cultivating mindfulness, even small steps taken in daily life are significant. They are significant not because on their own they will change the whole world. They are significant because they contribute to a network of causes and conditions that can change the whole world. Be it living more simply, shifting to a vegetarian or vegan diet, recycling, forgoing unnecessary travel by car or plane, or even acquiring an electric car, these deeds become meaningful not because the world will change if one individual acts in this way. They are meaningful because they embody awareness of the global crisis and express it on the individual level as a form of training in mindfulness and ethical responsibility.

Of course, the more who act in this way, the greater the effects will be. This ties in with the internal and external dimensions of mindfulness, where the internal builds the foundation for the external. It is precisely through embodying what needs to be done on the personal level that the outside world can be positively affected.

By training oneself to face the crisis with mindful balance, one will be able to exemplify mindfulness in an authentic way and share this attitude with others, inspiring them to cultivate the same. Equipped with this attitude, any ecological activism to confront the crisis has the greatest potential for success.

Concentration

Different definitions of right concentration can be found in the early discourses, which either list the four absorptions or else stipulate unification of the mind cultivated in conjunction with the other seven path factors. Closer comparative study shows the latter definition to be quite probably the earlier one of the two (Anālayo 2019b). This can take the following form:[34]

> Right view, right intention, right speech, right action, right livelihood, right effort, and right mindfulness; if based on arousing these seven factors, on being supported [by them] and equipped [with them], the mind progresses well and attains unification, then this is reckoned noble right concentration with its arousing, with its supports, and with its equipment.

In the meditative approach presented here, such unification of the mind would find its implementation in the cultivation of compassion as a boundless radiation, discussed in the previous chapter.

The Middle Path

In the Discourse on Turning the Wheel of Dharma, the teaching of the eightfold path serves to exemplify a middle-path approach. The principle underlying this middle path is as follows:[35]

> Five monastics, you should know that there are two extreme undertakings that those who are on the path should not practice: the first is attachment to sensual pleasures, which is a lowly act, undertaken by the ordinary person; the second is to torture oneself and [make] oneself suffer,

---

[34] MĀ 189 at T I 735c5, parallel to MN 117 at MN III 71,16 and Up 6080 at D 4094 *nyu* 44a2 or P 5595 *thu* 83b1.

[35] MĀ 204 at T I 777c26. The parallel MN 26 does not cover the teaching of the middle path, which is, however, found in SN 56.11 at SN V 421,2 (translated by Bodhi 2000: 1844); for a comparative study and a translation of MĀ 204 see Anālayo 2011a and 2011b: 170–189.

which is an ignoble condition and not connected to what is beneficial.[36]

Five monastics, abandon these two extremes and take up the middle path, which accomplishes understanding, accomplishes wisdom, and accomplishes [inner] certainty and the attainment of mastery, and which leads to wisdom, leads to awakening, and leads to Nirvana, namely the eight-[fold] right path, from right view to right concentration.

The "five monastics", mentioned here, had been companions of the Buddha when he was still in search of awakening and was undertaking ascetic practices (Anālayo 2017d: 51–76). After having pursued these for some time, he realized that they were not conducive to liberation. When he abandoned asceticism, these five companions left him, in the belief that he had thereby abandoned the path to awakening. This was not the case, however, as he had only changed his approach.

At the present juncture of events, after having reached awakening, the Buddha had come to meet these five former companions in order to share his discovery with them. Due to their belief that he had given up the path to awakening, they were naturally disinclined to believe his claim. For this reason, he had to clarify at the outset the nature of the path that had led him to realization.

The notion of a middle path between self-torment and indulgence can conveniently be related to facing climate change. As noted by Daniels (2010b: 966): "one major theme imbued in the Eightfold Path, of particular relevance to sustainability issues, is the principle of moderation or the 'Middle Way'." The Karmapa (2009: 82) reasons, regarding "the key principle

---

[36] The translation is based on adopting a variant reading that dispenses with an additional reference to "searching".

of the Middle Way", that "our lifestyle today should be modelled on this principle—neither too hard nor too indulgent."

In other words, the situation clearly demands action to be taken on the personal level in line with the intention for renunciation. Certain comforts and indulgences need to be abandoned in order to alleviate the burden on the environment and truthfully embody the commitment to maintaining living conditions on the earth. At the same time, however, this should not lead to self-torture by going to extremes. Here mindfulness has an important role to fulfill by monitoring and providing the required feedback, noting whenever a lack of balance has occurred and calling for adjustment.

Another aspect of interest, related to the Buddha's meeting with his five former companions, is that along the way he had chanced across another religious practitioner. This practitioner was thus the first person to encounter the Buddha after he had reached awakening. According to the report of this meeting, the Buddha announced his plans in the following manner:[37]

> I am going to ... beat the sublime drum of the deathless and to turn the unsurpassable wheel of the Dharma.

The reference to turning the wheel of the Dharma concerns the impending delivery of his first teaching to his five former companions. The notion of beating the drum of the deathless reflects the successful completion of the Buddha's quest. According to the same discourse, this quest had been as follows:[38]

> Formerly, when I had not yet awakened to supreme, right, and complete awakening, I also had this reflection: "I am

---

[37] MĀ 204 at T I 777b26, parallel to MN 26 at MN I 171,11 (translated by Ñāṇamoli 1995/2005: 263).
[38] MĀ 204 at T I 776a26, parallel to MN 26 at MN I 163,15.

actually subject to disease myself and I naïvely search for what is subject to disease; I am actually subject to old age, subject to death, subject to worry and sadness, and subject to defilement myself and I naïvely search for what is subject to defilement.

What if I now rather search for the supreme peace of Nirvana, which is free from disease, search for the supreme peace of Nirvana, which is free from old age, free from death, free from worry and sadness, and free from defilement?"

The notion of the deathless can serve to summarize the Buddha's pre-awakening quest for liberation, described in the present extract. Here the deathless does not refer to the achievement of a state of eternal life. Instead, it signifies a liberating insight which results in the complete conquest of any fear of death.

According to early Buddhist thought, with full awakening gained, the Buddha had reached a condition of the mind that was completely composed even when faced with his own passing away. The same holds for those of his disciples who had also become completely liberated from defilements. Their conquest of death neither avoids the passing away of their physical bodies nor leads to gaining a state of immortality in a heavenly realm. Instead, it involves a supreme condition of freedom of the mind, such that even the terror of mortality has completely lost its sting.

Mindfulness of Death

A meditation practice of direct relevance to this notion of the deathless, as a freedom from the terror of mortality, is mindfulness of death. This can take the following form:[39]

---

[39] EĀ 40.8 at T II 742a25, parallel to AN 6.19 at AN III 306,6 and AN 8.73 at AN IV 319,23 (translated by Bodhi 2012: 878 and 1221); see

> One gives attention to the perception of death, collecting
> mindfulness to the fore, with a mind that is unshaken, be-
> ing mindful of the exhalation and the inhalation for the
> time it takes for them to go out and return.

Although the instruction is fairly simple, putting it into practice can be challenging. This is in considerable part because
our modern society has become so accustomed to avoiding the
fact of death.

The different defense mechanisms employed to ignore mortality, both one's own and that of others, have been studied in
detail in clinical psychology. A range of publications are available on what at times comes under the header of "Terror Management Theory" (Greenberg et al. 1986). This is the *theory* that
explains how human beings *manage* their existential *terror*.

Human beings share with animals the instinct for self-preservation. In the case of humans, this instinct combines with
the awareness that death is unavoidable. The combination of
the instinctive drive for self-preservation and the knowledge of
the inevitability of death creates the potential for paralyzing
terror. As soon as death comes within the range of attention,
human beings tend to react with various defense mechanisms.
The most common one is trying to distract oneself.

Should distraction not suffice, denial ensues. This can take
two forms. In one form of denial, death is pushed far away into the distant future. The acknowledgement that one is indeed
going to die is accompanied by the reassuring assumption that
this will only happen after a very long time. At that far away
time in the distant future, one will come back to this issue, but

---

also Anālayo 2016d: 200–207 and 2018b: 90–95. EĀ 40.8 also brings
in the cultivation of the awakening factors, but these are not mentioned in the two Pāli parallels.

not now. After all, there are more important things to do at present. No need to be too concerned about death at this time, as it is still so far away.

The other form of denial pretends that somehow, in a way not further specified, one is exempt from mortality. Other people are indeed mortal, this can hardly be denied, but somehow, in some way, oneself is not really subject to the same. Death is out there, but not in here.

These two modalities of denial could be summarized as "not now" and "not me". They can be countered by repeatedly directing mindfulness to the recognition that one's own death can happen at any time. Mortality is actually the birthday present every human being receives right on coming into existence. Although this should be obvious, it takes much courage and effort to face what most individuals shy away from: Death is certain.

As a consequence of being made aware of their mortality, people can cling strongly to their views and sense of identity as a way of fending off the sense of being threatened. Just being briefly reminded of the fact of death can make individuals react in ways that are more narrow-minded, biased, and fundamentalist, as strategies for avoiding the realization of their own mortality.

Recollection of death serves to counter these tendencies and to clarify priorities in life. In the face of mortality, how should life be lived in such a way that there will be no regrets at the time of death?

Although at first sight this might seem paradoxical, through regular recollection of death one becomes much more alive. One becomes more alive to the opportunities of the present moment, to the importance of making the best use of it instead of squandering it in meaningless activities.

In this way, mindfulness of death can have a remarkable impact on one's priorities and relationships; it can positively change one's whole life. By allowing death to become part of one's life, a process of becoming complete and whole can take place.

Such mental cultivation is of additional relevance to facing the current climate emergency. It may well be the same defense mechanisms used to ward off recognition of mortality that contribute to denying or ignoring climate change. The same tendency manifests in clinging strongly to views and identities in order to fend off the feeling of being threatened.

Two strategies of psychological distancing in the face of climate change correspond closely to the two modalities of denying mortality by way of "not me" and "not now". According to McDonald et al. (2015: 113), "confronted by evidence for a serious threat such as climate change, people may be motivated to see the location of impacts as far from themselves" in a geographical sense. Moreover, "even if one accepts that climate change is happening, and that the impacts will be severe, one might still feel psychologically distant from climate change because the potential impacts are a long way in the future" (2015: 112).

In view of this, mindfulness of death can make a rather substantial contribution to the challenge of facing climate change. Countering the two strategies of denial in the form of "not me" and "not now" is as relevant to climate change as it is to mortality. The dire effects of climate disaster can strike anywhere in the world, at any time.

Summary
The vision of a complete destruction of the whole earth has a soteriological potential in early Buddhist thought, by way of

driving home the truth of impermanen(
go. The same truth applies to oneself;
one's own mortality is the internal (
facing climate catastrophe on the ex'
through mindfulness practice the challenge
can become a path to awakening. This can take the form of
middle path aloof from the two extremes of apathy and agita-
tion, and aloof from indulgence and self-tormenting.

The actual implementation of this middle path has the four
noble truths as its guiding principles, ethical living as its support, and mindfulness as its constant companion, monitoring
what is taking place. Intentions of renunciation, non-harm, and
non-ill will keep the mind on track as it invests effort into confronting the three root defilements on the internal and external
levels. In relation to verbal activity, the truth of climate change
and its repercussions need to be communicated with balance, a
balance that also informs environmental activism and one's personal lifestyle adjusted in such a manner as to minimize one's
carbon footprint.

In this way, every step taken along this path can serve to diminish pollution both without and within.

Conclusion

The cultivation of mindfulness facilitates approaching the disastrous environmental repercussions caused by the influence of the three root defilements without succumbing to them oneself. The crisis itself can be seen as the result of these three mental defilements, in particular rampant greed and the deluded tendency to ignore its repercussions. Although anger may at first sight seem less prominent, with the imminent deterioration in living conditions it can safely be expected to become more conspicuous.

The cultivation of mindfulness to face climate change rests on the compassionate intention for the absence of harm. It monitors and finetunes the contribution made by compassion, ensuring that one neither succumbs to its near enemy of grief nor switches off due to being unable to face it any longer. Viewed from this perspective, facing climate change becomes a mindfulness practice all the way through. Not only that, but its final goal is precisely a raising of the level of mindfulness on a global scale.

The potential of mindfulness in this respect can nowadays be tapped more easily due to its worldwide spread, as a result of having been adopted in a variety of areas in contemporary society and modern culture (Wilson 2014). Training in mindfulness has become available throughout the world and is accessible to people from a wide variety of backgrounds.

This is the other side of the coin of the present situation. With all its catastrophic dimensions, it is at the same time also an outstanding opportunity. It is an opportunity to step out of detrimental patterns ingrained in human civilization as it is at present and move to a level of interaction among human beings that gives precedence to the common welfare over the individual benefit. The challenge posed by the crisis, if handled with mindfulness and compassion on a broad scale, can become an occasion to learn to work together to maintain the living conditions required for the survival of human civilization. Keeping in mind the positive view of this potential will be a crucial asset in facing any adversities.

Working together to ensure sustainability of life on earth indubitably requires stepping out of the narrow confines of self-centeredness, based on rigidly held racial, political, religious, and social identities. At this stage, it is no longer possible to privilege the individual over the communal, the regional over the national, and the national over the international. Instead, human beings all over the world must come to appreciate what they all have in common, the potential to become what so far they have not yet really become: *homo sapiens sapiens*, truly "wise" human beings.

Appendix: Meditation

The material in the preceding chapters is meant to help strengthen mental resilience for facing the current environmental challenge and for taking appropriate action. Each of the four chapters provides a background to a particular aspect of the meditation presented here.

In the case of the first chapter, this is contemplation of the earth element. Such contemplation can make it a matter of personal experience that one is intrinsically related to nature outside. The second chapter relates to contemplation of the mind, with a particular emphasis on recognition of the three root defilements. The third chapter concerns the cultivation of compassion. The last chapter introduces contemplation of impermanence on a global scale and recollection of death. Coming to terms with one's own mortality has its complement in the ability to face the possible end of human civilization with inner balance, an indispensable foundation for taking appropriate action.

Although inspired by the early discourses, the actual instructions summarized below and also offered as guided instructions online are my own and come with no claim of being accurate reflections of meditation practice undertaken in ancient India.

The basic suggestion would be to proceed through these meditations step by step. A convenient way of doing so could be during a four-week period, finding time each weekend to study one chapter and during the ensuing week to practice the corresponding meditation on a daily basis.

The meditations presented here are based on extracts from more comprehensive practices that have considerably more to offer. The combination of these extracts yields a viable form of meditation that fits the challenge of facing climate change and at the same time has a liberating potential. Nevertheless, in addition to cultivating such meditation on a daily basis, it would be opportune to explore the relevant complete practices during a time of retreat.

Building on the daily practice of the meditation presented here, a time of the year set apart for silent retreat could be used to cultivate contemplation of all four elements in the way described in the discourse (Anālayo 2018b: 65–81), even to explore the full scheme of the four establishments of mindfulness. Other opportunities for deepening the practice would be to implement the complete instructions on mindfulness of breathing in sixteen steps (Anālayo 2019e: 6–154) or else to develop all of the four divine abodes (Anālayo 2015b: 154–162). Such full exploration will provide additional depth to the meditation presented here and strengthen its daily practice.

Contemplation of the Earth

In order to cultivate insight into the close relationship between the body and the earth on a personal level, meditation practice can begin with a focus on the earth element. Although such an approach is inspired by the contemplation of the four elements described in the Discourse on the Establishments of Mindfulness, it lacks the analytical edge of this exercise, due to taking up only one of the four elements.

For exploring the earth element in daily meditation, the recommended approach is to use a body scan. Such a form of practice is not described in the early discourses and only seems to have come into existence in later times (Anālayo 2020b). Its purpose is to provide a grounding in bodily presence, making it easy to collect the mind and avoid that it succumbs to distraction.

In actual practice, after having taken a moment to settle in by just being with the presence of the body in the sitting posture, the scan can begin with the head and from there proceed to the neck, shoulders, arms, hands, torso, hips, legs, and feet. At first it might be preferable to take the limbs separately, but eventually these can be covered at the same time.

The meditative task is simply to be aware of a particular part of the body in the knowledge that there is the internal earth element, in the sense that there is some degree of solidity in this part of the body. When cultivating this meditation there is no need to strain in order to feel distinctly and with total precision the presence of solidity in each and every part of the body. Obviously, the body is solid; there is solidity in each part of the body. Given that this is already clear, it is not necessary to struggle in order to prove that. It suffices just being aware of the body and knowing that there is solidity, which is sensed only to the degree to which this naturally manifests.

Having completed the body scan can then lead on to sensing the solidity of the ground below, wherever this is in direct contact with the body. This is a way of transitioning to becoming aware of the external earth element. Even sitting on a chair on the highest floor of a skyscraper, there definitely is solidity below that reaches all the way down to the earth. Awareness can note the sense of gravity and allow the body to relax into that gravitational pull, letting all bodily and mental tension sink into the ground.

Having in this way come into contact with the earth below can lead over to a perception of the extensiveness of the earth in all directions. This can be done by first becoming aware of the frontal direction, in the acknowledgement that the element of solidity felt below the body extends to the front into the far distance. Proceeding from the front to the right, then the back, and finally the left, eventually a perception of the vastness of the external earth element can arise. The body can feel firmly grounded in this experience of solidity that extends into all directions; in fact, it is an integral part of it.

The perception of the body as an integral part of the earth can be further strengthened by way of the following reflection, brought in briefly and only to the extent to which this supports the meditation, without leading to mental chatter: The body depends on a constant supply of solid food and is thereby in a relationship of exchange with the external earth element. It is entirely dependent on the earth for its survival. The same holds not only for the earth element. The body also needs the water element in the form of beverages, the fire element in the form of protection from extremes of temperature, and the wind element in the form of breathing. The oxygen breathed in comes from plants that live on the surface of the earth and in the ocean.

The last reflection can lead over to becoming aware of the process of breathing. Ideally this is done while maintaining whole-body awareness. In other words, instead of cultivating an exclusive focus on the breath, the process of breathing can be experienced as part of the whole body seated on the earth. With every breath, an exchange takes place with the plants on the earth. For the body's survival, this is even more vital than its food supply.

The practice of the body scan can serve as a convenient tool for adjusting to the degree of distraction of the mind. When the mind tends to wander frequently, repeated and swift scanning can help to counter this tendency. Once, sooner or later, the mind becomes willing to settle down, the time has come to give attention to the breath. Throughout, the relationship between the body and the earth, the dependency of the body on the earth for its survival, can remain as the central theme.

Contemplation of the Mind

Building on some experience with contemplation of the earth element, the first moment of just settling in by becoming aware of the presence of the body in the sitting posture can also serve as a moment to turn awareness to the present condition of the mind. This just requires checking in to see where the mind is at present and taking a moment to sense fully its texture and condition, in particular its distinct quality when mindfulness is present.

The awareness of the mind established in this way can then become a companion during the body scan or when observing the breath. This can take the form of meta-awareness while undertaking these practices. The difference compared to the practice done earlier is similar to the difference between just reading these lines and reading them while being aware of the fact that one is reading.

Learning to keep an eye on the mind in this way will considerably strengthen one's ability to avoid getting caught up in daydreams or fantasies. Nevertheless, sooner or later some distraction or the other is bound to happen.

Whenever the mind gets off track and this is noticed, it is of utmost importance to let go right away of any frustration or negativity. It is simply the nature of the mind to wander; there is nothing surprising in this. Instead of getting upset, noting that some daydream or fantasy has taken the mind for a ride can be recognized with an inner smile in the knowledge that this is just the way of the mind.

Recognizing the occurrence of a distraction can serve as a welcome opportunity for exploring contemplation of the mind in more detail. This can take the form of discerning, first of all, the predominant feeling tone, the affective quality of what just happened in the mind. If this has been pleasant, chances are that the mental wandering was related to greed. If it has been unpleasant, chances are that it was related to anger and aversion. If there were neutral feeling tones, this can be a signal for a deluded state of mind.

Needless to say, greed and anger are also manifestations of delusion. But in the present context the label "delusion" can conveniently be employed for those distractions that do not fit either of the other two categories well. This is when the mind is just ambling around with no purpose, without being in an obvious state of either greed or anger.

The identification of the nature of the mental wandering that just occurred can come together with a recognition of the condition of the mind right now, when mindfulness is present. Compared to its earlier distracted condition, the mind has become so much more open, aware, alive, and receptive. Taking time to savor the different actual condition of the mind makes it obvious why a mind relatively free from the influence of the three root defilements is indeed preferable.

Turning to the present condition of the mind can also reveal the actual feeling tone present when mindfulness is established. Close inspection uncovers the presence of a very subtle type of pleasant feeling tone: the joy of being in the here and now. Keeping attuned to this wholesome type of joy will further strengthen the mind's ability to stay on track. This in a way reflects the potential of mindfulness to lead beyond grief and sadness, mentioned earlier.

Cultivating Compassion

The presence of the wholesome joy of being in the present moment provides an ideal foundation for the arousing of compassion or the other divine abodes. Such arousing can be undertaken based on employing either a particular phrase that expresses the sentiment of the relevant divine abode or else by using a mental image or picture. In the case of compassion, the sentiment would be the wish for non-harm, and an image could be anything that arouses within oneself the mental attitude of compassion. Any image that fulfils this purpose can serve as the starting point.

In case one decides to start with mental reflections in the form of phrases, then these are best kept short and concise right from the outset. Both reflections and mental images can be skillful means to arouse compassion; once they have fulfilled this task, however, they can be left behind. Such tools are supports meant to lead on to a stage of practice where they are no longer needed.

Once they seem no longer necessary, a shift can take place from *doing* compassion to simply *being* compassion, by way of allowing one's whole body and mind to be suffused with compassion.

After having dwelled for some time in this experience, the compassion can be allowed to radiate in the different directions. Such radiation can begin by pervading the front, then the right, the back, and the left; this is similar to the earlier practice of contemplating the external earth element. Having in this way established pervasion of the four directions the radiation of compassion can also extend upwards and then downwards.

The practice of the radiation could be compared to a source of light that is surrounded by a curtain on all sides. To allow the light to shine in all directions, one slowly and gently pulls away the curtain. No need to push or exert force in any way. It does not matter how far the light of compassion is able to shine right now. The task is only one of removing any boundary, of allowing the mind to become naturally boundless. With continued practice, at times the inner light will shine with increasing strength and its illumination will spread far and wide. However short or far it may spread, a temporary liberation of the mind is reached as soon as the radiation has become boundless in all directions.

With the radiation established in all directions, one simply remains in the spaciousness of the mental liberation by compassion, without paying further attention to individual directions. In terms of the light simile, the curtains having been gently pulled away, now the light just keeps shining in all directions. A soft awareness of the body in the sitting posture and of the continuity of breathing can serve as stabilizers of the mind. For the same purpose, it can be helpful to use the times of inhalations for attending to the felt sense of compassion and the times of exhalations to being more aware of the spaciousness of the mind.

Within such spaciousness of the mind, any defilement has no chance to remain. As soon as a distraction is noticed, one just allows the mind to return to its condition of spaciousness and the mental narrowness that accompanies a defilement evaporates on its own.

Contemplation of Impermanence

Out of the full scheme of instructions on mindfulness of breathing, for the present context the recommendation is to rely on its last four steps, which combine awareness of inhalations and exhalations with contemplation of impermanence, dispassion, cessation, and letting go.

Awareness of the process of breathing, established after the body scan for experiencing the internal earth element, can now be invested with an emphasis on directly experiencing its impermanent nature. In terms of one's mental attitude, this involves a shift from the compassion cultivated earlier to equanimity now becoming predominant, with the understanding that the breath is nothing but change. Just as the breath, so everything else is also impermanent.

Those who like to work with more detail could bring in the five aggregates, discussed in the first chapter (see above p. 21). The body within which the breath is experienced corresponds to the first aggregate of bodily form. The sensations caused by the breath belong to the second aggregate of feeling tones. The discerning of inhalations and exhalations relies on the third aggregate of perception. The intention to stay with the breath and return to it whenever the mind has become distracted is the fourth aggregate of volitional formations. The knowing of all these aspects of the present moment's experience of the breath is consciousness.

Contemplation can take up each of these five aggregates one after the other, with an emphasis on their impermanent nature. Eventually, they can be combined in a comprehensive appreciation of the changing nature of all aspects of subjective experience.

Once insight into impermanence has been well established, in whichever way this has been done, it can lead over to cultivating dispassion. This takes place by letting the implications of the changing nature of all phenomena transform one's affective attitude toward them and diminish one's clinging.

The more dispassion grows, the easier it will become to be at ease with the ending of things, with their cessation. Depending on the present situation and one's overall preferences, the contemplation of cessation can be done either briefly or else in more depth.

Exploring the step of cessation can take place by taking up the thought of one's own death, the recognition that eventually this body will be bereft of life and the breath will stop. The thought of one's own death can be brought up either as a brief reminder, or else it can be contemplated in some detail, according to what feels appropriate at that time. For a more comprehensive reflection, the same practice can be extended to the whole earth, with the awareness that at some time in the future it will be completely destroyed. All traces of human civilization will come to an end. Nothing will remain forever.

The thought of one's own death can similarly be related to the alternation between inhalations and exhalations. With every inhalation, there can be an emphasis on the fact that, in principle, this could be the last breath. With every exhalation, one relaxes and lets go.

Correlating these two aspects of recognizing mortality and letting go to the inhalations and exhalations respectively enables adjusting the practice as needed. When the mind tries to dismiss the fact of death, more emphasis could be given to the inhalations. This is not by way of changing the inhalations in any way, but only in the sense of giving more emphasis to the corresponding reflection or perception. Breathing remains natural throughout. When the truth of mortality becomes too agitating, more emphasis can be given to relaxing and letting go with the exhalations.

After having been related to the exhalations, letting go can then become a continuous theme of the meditation. As the last of the four insight themes, letting go completes the trajectory from impermanence to dispassion and cessation. Such letting go can be aimed in particular at any manifestation, however subtle it may be, of the three root defilements of greed, hatred, and delusion. Cultivating increasing degrees of inner freedom from the three root defilements enables facing their external manifestations with maximal effectiveness. In this way, deepening meditative insight on the internal level builds the required foundation for countering the environmental destruction in the world outside to the best of one's ability.

Summary

The meditation practice described above is meant to offer a relation to each of the four establishments of mindfulness and at the same time incorporate both insight and tranquility.

The contemplation of the internal earth element takes up one of the four elements, described in the Discourse on the Establishment of Mindfulness under the header of contemplation of the body. The same practice also incorporates mindfulness of the process of breathing, experienced as another dimension of the relationship between this body and the earth.

Contemplation of the mind in the way described above corresponds to the first three mental states listed in the Discourse on the Establishment of Mindfulness. The meditation presented here covers also contemplation of feeling tones, by way of attending to the affective quality of the mind when a distraction had been present as well as through noting the pleasant feeling tone of a mind that is undistracted.

Taking off from the pleasant feeling tone of being in the present moment, the meditation moves on to cultivation of compassion as a boundless radiation, thereby ensuring that the formal development of meditative tranquility complements the emphasis on insight in the remainder of the meditative approach presented here.

In order to implement the fourth establishment of mindfulness, contemplation of dharmas, the instructions for the last tetrad of mindfulness of breathing provide a helpful template. The basic progression involves contemplating impermanence, dispassion, cessation, and letting go (Anālayo 2019e: 100–119).

The third of these four themes, cessation, provides an opportunity to implement recollection of death, in terms of one's own mortality as well as the ultimate death of the earth.

The meditation as a whole attempts to combine progress to awakening with the cultivation of qualities and insights directly related to climate change. It intends to make the relationship of the body to the earth a matter of direct personal experience. It also tries to provide tools to counter the influence of the three root defilements and to inculcate compassion as the crucial attitude of non-harm underlying environmental concerns. Learning to face death and the potential ending of human civilization with equanimity is perhaps the most crucial contribution to developing the mental resilience needed for facing the dire consequences of climate change.

Abbreviations

AN	*Aṅguttara-nikāya*
CBETA	Chinese Buddhist Electronic Text Association
D	Derge edition
DĀ	*Dīrgha-āgama* (T 1)
Dhp	*Dhammapada*
DN	*Dīgha-nikāya*
EĀ	*Ekottarika-āgama* (T 125)
EĀ²	*Ekottarika-āgama* (T 150A)
Jā	*Jātaka*
MĀ	*Madhyama-āgama* (T 26)
MN	*Majjhima-nikāya*
Mp	*Manorathapūraṇī*
P	Peking edition
Ps	*Papañcasūdanī*
SĀ	*Saṃyukta-āgama* (T 99)
SĀ²	*Saṃyukta-āgama* (T 100)
SHT	Sanskrithandschriften aus den Turfanfunden
SN	*Saṃyutta-nikāya*
Sn	*Sutta-nipāta*
T	Taishō edition (CBETA)
Th	*Theragāthā*
Up	*Abhidharmakośopāyikā-ṭīkā*
Vin	*Vinaya*
Vism	*Visuddhimagga*
[ ]	supplementation

References

Allon, Mark and B. Silverlock 2017: "Sūtras in the Senior Kharoṣṭhī Manuscript Collection with Parallels in the Majjhima-nikāya and/or the Madhyama-āgama", in *Research on the Madhyama-āgama*, Dhammadinnā (ed.), 1–54, Taipei: Dharma Drum Publishing Corporation.

Almiron, Núria and J. Xifra (ed.) 2020: *Climate Change Denial and Public Relations, Strategic Communication and Interest Groups in Climate Action*, New York: Routledge.

Amaro, Ajahn 2013: *For the Love of the World*, Great Gaddesden: Amaravati Publications.

Anālayo 2003: *Satipaṭṭhāna, the Direct Path to Realization*, Birmingham: Windhorse Publications.[1]

— 2006: "The Ekottarika-āgama Parallel to the Saccavibhaṅga-sutta and the Four (Noble) Truths", *Buddhist Studies Review*, 23.2: 145–153 (reprinted in 2016a).

— 2009a: "Karma and Liberation — The Karajakāya-sutta (AN 10.208) in the Light of its Parallels", in *Pāsādikadānaṃ, Festschrift für Bhikkhu Pāsādika*, M. Straube, R. Steiner, J. Soni, M. Hahn, and M. Demoto (ed.), 1–24, Marburg: Indica et Tibetica (reprinted in 2012c).

— 2009b: "Vimutti", in *Encyclopaedia of Buddhism*, W.G. Weeraratne (ed.), 8.3: 615–622, Sri Lanka: Department of Buddhist Affairs.

— 2009c: "Virāga", in *Encyclopaedia of Buddhism*, W.G.

---

[1] Most of my publications listed here can be downloaded for free at: https://www.buddhistinquiry.org/resources/publications-by-bhikkhu-analayo.

Weeraratne (ed.), 8.3: 688–690, Sri Lanka: Department of Buddhist Affairs.

— 2010: "The Mahācattārīsaka-sutta in the Light of its Parallels: Tracing the Beginnings of Abhidharmic Thought", *Journal of the Centre for Buddhist Studies, Sri Lanka*, 8: 59–93 (reprinted in 2012c).

— 2011a: "Brahmā's Invitation, The Ariyapariyesanā-sutta in the Light of its Madhyama-āgama Parallel", *Journal of the Oxford Centre for Buddhist Studies*, 1: 12–38 (reprinted in 2012c).

— 2011b: *A Comparative Study of the Majjhima-nikāya*, Taiwan: Dharma Drum Publishing Corporation.

— 2011c: "Right View and the Scheme of the Four Truths in Early Buddhism, The Saṃyukta-āgama Parallel to the Sammādiṭṭhi-sutta and the Simile of the Four Skills of a Physician", *Canadian Journal of Buddhist Studies*, 7: 11–44 (reprinted in 2015d).

— 2012a: "The Case of Sudinna: On the Function of Vinaya Narrative, Based on a Comparative Study of the Background Narration to the First Pārājika Rule", *Journal of Buddhist Ethics*, 19: 396–438 (reprinted in 2017f).

— 2012b: "The Chinese Parallels to the Dhammacakkappavattana-sutta (1)", *Journal of the Oxford Centre for Buddhist Studies*, 3: 12–46 (reprinted in 2015d).

— 2012c: *Madhyama-āgama Studies*, Taipei: Dharma Drum Publishing Corporation.

— 2012d: "Protecting Oneself and Others Through Mindfulness — The Acrobat Simile in the Saṃyukta-āgama", *Sri Lanka International Journal of Buddhist Studies*, 2: 1–23 (reprinted in 2015d).

— 2013a: "The Chinese Parallels to the Dhammacakkappavattana-sutta (2)", *Journal of the Oxford Centre for Buddhist Studies*, 5: 9–41 (reprinted in 2016a).

— 2013b: *Perspectives on Satipaṭṭhāna*, Cambridge: Windhorse Publications.

— 2014a: "Maitreya and the Wheel-turning King", *Asian Literature and Translation: A Journal of Religion and Culture*, 2.7: 1–29 (reprinted in 2017a).

— 2014b: "The Mass Suicide of Monks in Discourse and Vinaya Literature", *Journal of the Oxford Centre for Buddhist Studies*, 7: 11–55 (reprinted in 2017f).

— 2014c: "On the Five Aggregates (4) — A Translation of Saṃyukta-āgama Discourses 33 to 58", *Dharma Drum Journal of Buddhist Studies*, 14: 1–71

— 2014/2015: "Discourse Merger in the Ekottarika-āgama (2), The Parallels to the Kakacūpama-sutta and the Alagaddūpama-sutta", *Journal of Buddhist Studies, Sri Lanka*, 12: 63–90 (reprinted in 2016).

— 2015a: "Compassion in the Āgamas and Nikāyas", *Dharma Drum Journal of Buddhist Studies*, 16: 1–30.

— 2015b: *Compassion and Emptiness in Early Buddhist Meditation*, Cambridge: Windhorse Publications.

— 2015c: "Discourse Merger in the Ekottarika-āgama (1), The Parallel to the Bhaddāli-sutta and the Latukikopama-sutta, Together with Notes on the Chinese Translation of the Collection", *Singaporean Journal of Buddhist Studies*, 2: 5–35.

— 2015d: *Saṃyukta-āgama Studies*, Taipei: Dharma Drum Publishing Corporation.

— 2016a: *Ekottarika-āgama Studies*, Taipei: Dharma Drum Publishing Corporation.

— 2016b: *The Foundation History of the Nuns' Order*, Bochum: Projektverlag.

— 2016c: "The Legal Consequences of pārājika", *Sri Lanka International Journal of Buddhist Studies*, 5: 1–22.

— 2016d: *Mindfully Facing Disease and Death, Compassionate Advice from Early Buddhist Texts*, Cambridge: Windhorse Publications.

— 2016e: "The Vessantara-Jātaka and Mūlasarvāstivāda Vinaya Narrative", *Journal of the Oxford Centre for Buddhist Studies*, 11: 11–37 (reprinted in 2017f).

— 2017a: *Dīrgha-āgama Studies*, Taipei: Dharma Drum Publishing Corporation.

— 2017b: *Early Buddhist Meditation Studies*, Barre: Barre Center for Buddhist Studies.

— 2017c: "How Compassion Became Painful", *Journal of the Centre for Buddhist Studies, Sri Lanka*, 14: 85–113.

— 2017d: *A Meditator's Life of the Buddha, Based on the Early Discourses*, Cambridge: Windhorse Publications.

— 2017e: "Some Renditions of the Term Tathāgata in the Chinese Āgamas", *Annual Report of the International Research Institute for Advanced Buddhology at Soka University*, 20: 11–21.

— 2017f: *Vinaya Studies*, Taipei: Dharma Drum Publishing Corporation.

— 2018a: *Rebirth in Early Buddhism and Current Research*, Boston: Wisdom Publications.

— 2018b: *Satipaṭṭhāna Meditation: A Practice Guide*, Cambridge: Windhorse Publications.

— 2019a: "Craving and dukkha", *Insight Journal*, 45: 35–42.

— 2019b: "Definitions of Right Concentration in Comparative Perspective", *Singaporean Journal of Buddhist Studies*, 5: 9–39.

— 2019c: "Immeasurable Meditations and Mindfulness", *Mindfulness*, 10: 2620–2628.

—— 2019d: "In the Seen Just the Seen: Mindfulness and the Construction of Experience", *Mindfulness*, 10: 179–184.

—— 2019e: *Mindfulness of Breathing, A Practice Guide and Translations*, Cambridge: Windhorse Publications.

—— 2019f: "Pārājika Does Not Necessarily Entail Expulsion", *Annual Report of the International Research Institute for Advanced Buddhology at Soka University*, 22: 3–8.

—— 2019g: "A Task for Mindfulness: Facing Climate Change", *Mindfulness*, 10: 1926–1935.

—— 2020a: "A Brief History of Buddhist Absorption", *Mindfulness*, 11.

—— 2020b: "Buddhist Antecedents to the Body Scan Meditation", *Mindfulness*, 11.

—— 2020c: "External Mindfulness", *Mindfulness*, 11.

Anthony, Katey Walter, T. Schneider von Deimling, I. Nitze, S. Frolking, A. Emond, R. Daanen, P. Anthony, P. Lindgren, B. Jones, and G. Grosse 2018: "21st-century Modeled Permafrost Carbon Emissions Accelerated by Abrupt Thaw Beneath Lakes", *Nature Communications*, 9. 3263: 1–11.

Aronson, Harvey B. 1980/1986: *Love and Sympathy in Theravāda Buddhism*, Delhi: Motilal Banarsidass.

Bapat, P.V. 1957: "Atta-dīpa in Pali Literature", in *Liebenthal Festschrift, Sino-Indian Studies, Volume V Parts 3 & 4*, K. Roy (ed.), 11–13, Santiniketan: Visvabharati.

Bareau, André 1991: "Les agissements de Devadatta selon les chapitres relatifs au schisme dans les divers Vinayapiṭaka", *Bulletin de l'École Française d'Extrême Orient*, 78: 87–132.

Bechert, Heinz and K. Wille 1989: *Sanskrithandschriften aus den Turfanfunden, Teil 6*, Stuttgart: Franz Steiner.

—— 2004: *Sanskrithandschriften aus den Turfanfunden, Teil 9*, Wiesbaden: Franz Steiner.

Bernhard, Franz 1965: *Udānavarga*, Göttingen: Vandenhoeck & Ruprecht.

Bodhi, Bhikkhu 2000: *The Connected Discourses of the Buddha, A New Translation of the Saṃyutta Nikāya*, Boston: Wisdom Publications.

— 2009: "The Voice of the Golden Goose", in *A Buddhist Response to the Climate Emergency*, J. Stanley, D. R. Loy, and G. Dorje (ed.), 157–171, Boston: Wisdom Publications.

— 2012: *The Numerical Discourses of the Buddha, A Translation of the Aṅguttara Nikāya*, Boston: Wisdom Publications.

Brand, Uwe, R. Posenato, R. Came, H. Affek, L. Angiolini, K. Azmy, and E. Farabegoli 2012: "The End-Permian Mass Extinction: A Rapid Volcanic CO_2 and CH_4-climatic Catastrophe", *Chemical Geology*, 322: 121–144.

Brand, Uwe, N. Blamey, C. Garbelli, E. Griesshaber, R. Posenato, L. Angiolini, K. Azmy, E. Farabegoli, and R. Came 2016: "Methane Hydrate: Killer Cause of Earth's Greatest Mass Extinction", *Palaeoworld*, 25: 496–507.

Brough, John 1962/2001: *The Gāndhārī Dharmapada, Edited with an Introduction and Commentary*, Delhi: Motilal Banarsidass.

Buchanan, Tony W., S. L. Bagley, R. B. Stansfield, and S. D. Preston 2012: "The Emphatic, Physiological Resonance of Stress", *Social Neuroscience*, 7: 191–201.

Cassegård, Carl, L. Soneryd, H. Thörn, and A. Wettergren 2017: *Climate Action in a Globalizing World, Comparative Perspectives on Environmental Movements in the Global North*, New York: Routledge.

Chokyi Nyima Rinpoche 2009: "Very Dangerous Territory", in *A Buddhist Response to the Climate Emergency*, J. Stanley, D. R. Loy, and G. Dorje (ed.), 137–139, Boston: Wisdom Publications.

Collins, Steven 1987: "Kalyāṇamitta and Kalyāṇamittatā", *Journal of the Pali Text Society*, 11: 51–72.

— 1998: *Nirvana and Other Buddhist Felicities, Utopias of the Pali Imaginaire*, Cambridge: Cambridge University Press.

Cone, Margaret 1989: "Patna Dharmapada", *Journal of the Pali Text Society*, 13: 101–217.

Coomaraswamy, Ananda K. 1943: "Saṃvega, 'Aestethic Shock'", *Harvard Journal of Asiatic Studies*, 7.3: 174–179.

Cousins, L.S. 1996: "Good or Skilful? Kusala in Canon and Commentary", *Journal of Buddhist Ethics*, 3: 136–164.

Curtin, Deane 2017: "To Live as a Lotus among the Flames ... Buddhist Awakening in the Middle of the Climate Crisis", *Worldviews–Global Religions Culture and Ecology*, 21.1: 21–40.

Cutright, N. L., E. E. Padgett, S. R. Awada, J. M. Pabis, and L. D. Pittman 2019: "The Role of Mindfulness in Psychological Outcomes for Children Following Hurricane Exposure", *Mindfulness*, 10: 1760–1767.

Dalai Lama, Tenzin Gyatso 2009: "Mind, Heart, and Nature", in *A Buddhist Response to the Climate Emergency*, J. Stanley, D. R. Loy, and G. Dorje (ed.), 21–31, Boston: Wisdom Publications.

— 2011: *Beyond Religion: Ethics for a Whole World*, New York: Houghton Mifflin Harcourt.

Daniels, Peter L. 2010a: "Climate Change, Economics and Buddhism — Part I: An Integrated Environmental Analysis Framework", *Ecological Economics*, 69.5: 952–961.

— 2010b: Climate Change, Economics and Buddhism — Part 2: New Views and Practices for Sustainable World Economies", *Ecological Economics*, 69.5: 962–972.

Dahl, Cortland J., A. Lutz, and R. J. Davidson 2015: "Reconstructing and Deconstructing the Self: Cognitive Mechanisms in Meditation Practice", *Trends in Cognitive Sciences*, 19.9: 515–523.

Deeg, Max 1999: "The Saṅgha of Devadatta: Fiction and History of a Heresy in the Buddhist Tradition", *Journal of the International College for Advanced Buddhist Studies*, 2: 183–218.

Deese, R. S. 2019: *Climate Change and the Future of Democracy*, Cham: Springer.

DellaSalla, Dominick A. 2020: "Has Anthropocentrism Replaced Ecocentrism in Conservation?", in *Conservation, Integrating Social and Ecological Justice*, H. Kopnina and H. Washington (ed.), 91–104, Cham: Springer.

de Silva, Lily 2000: "Early Buddhist Attitudes Toward Nature", in *Dharma Rain, Sources of Buddhist Environmentalism*, S. Kaza and K. Kraft (ed.), 91–103, Boston: Shambala.

de Silva, Padmasiri 1990: "Buddhist Environmental Ethics", in *Dharma Gaia, A Harvest of Essays in Buddhism and Ecology*, A. H. Badiner (ed.), 14–19, Berkeley: Parallax Press.

Dhammadinnā, Bhikkhunī 2014: "Semantics of Wholesomeness: Purification of Intention and the Soteriological Function of the Immeasurables (appamāṇas) in Early Buddhist Thought", in *Proceedings of the International Conference 'Buddhist Meditative Traditions: Their Origin and Development'*, K. Chuang (ed.), 51–129, Taiwan: Dharma Drum Publishing Corporation.

— 2018a: "Discourses on the Establishment of Mindfulness (smṛtyupasthāna) quoted in Śamathadeva's Abhidharmakośopāyikā-ṭīkā", *Journal of Buddhist Studies, Sri Lanka*, 15: 23–38.

— 2018b: "When Womanhood Matters: Sex-essentialization and Pedagogical Dissonance in Buddhist Discourse", *Religions of South Asia*, 12.3: 274–313.

Dhammajoti, Bhikkhu K.L. 1995: *The Chinese Version of Dharmapada, Translated with Introduction and Annotations*, Sri Lanka: University of Kelaniya, Postgraduate Institute of Pali and Buddhist Studies.

— 2010: "The apramāṇa Meditation in the Sarvāstivāda, With Special Reference to maitrī-bhāvanā", *Journal of the Centre for Buddhist Studies, Sri Lanka*, 8: 165–186.

Dhammika, S. 2015/2018: *Nature and the Environment in Early Buddhism*, Kandy: Buddhist Publication Society.

Dietz, Siglinde 2007: "The Saptasūryodayasūtra", in *Indica et Tibetica 65, Festschrift für Michael Hahn zum 65. Geburtstag von Freunden und Schülern überreicht*, K. Klaus and J. U. Hartmann (ed.), 93–112, Wien: Arbeitskreis für tibetische und buddhistische Studien, Universität Wien.

Dumont, Louis 1962: "The Conception of Kingship in Ancient India", *Contributions to Indian Sociology*, 6: 48–77.

Dzigar Kongtrul Rinpoche 2009: "Minimum Needs and Maximum Contentment", in *A Buddhist Response to the Climate Emergency*, J. Stanley, D. R. Loy, and G. Dorje (ed.), 147–153, Boston: Wisdom Publications.

Eckel, Malcolm David 1997: "Is there a Buddhist Philosophy of Nature?", in *Buddhism and Ecology, The Interconnection of Dharma and Deeds*, M. E. Tucker and D. R. Williams (ed.), 327–349, Cambridge: Harvard University Press.

Filho, Walter Leal 2019a: "Introducing the International Climate Change Information Programme (ICCIP)", in *University Initiatives in Climate Change Mitigation and Adaptation*, W. L. Filho and R. L. Arcas (ed.), 3–11, Cham: Springer.

Filho, Walter Leal 2019b: "An Overview of the Challenges in Climate Change Communication Across Various Audiences, in *Addressing the Challenges in Communicating Climate*

Change Across Various Audiences, W. L. Filho, B. Lackner, and H. McGhie (ed.), 1–11, Cham: Springer.

Filho, Walter Leal, B. Lackner, and H. McGhie 2019: *Addressing the Challenges in Communicating Climate Change Across Various Audiences*, Cham: Springer.

Giustarini, Giuliano 2012: "The Role of Fear (Bhaya) in the Nikāyas and in the Abhidhamma", *Journal of Indian Philosophy*, 40.5: 511–531.

— 2018: "Healthcare in Pali Buddhism", *Journal of Religion and Health*, 57.4: 1224–1236.

Glass, Andrew and M. Allon 2007: *Four Gāndhārī Saṃyuktāgama Sūtras: Senior Kharoṣṭhī Fragment 5*, Seattle: University of Washington Press.

Gnoli, Raniero 1978: *The Gilgit Manuscript of the Saṅghabhedavastu, Being the 17th and Last Section of the Vinaya of the Mūlasarvāstivādin, vol. 2*, Rome: Istituto Italiano per il Medio ed Estremo Oriente.

Goldstein, Joseph 2009: "Except as We Have Loved, All News Arrives as from a Distant Land", in *A Buddhist Response to the Climate Emergency*, J. Stanley, D. R. Loy, and G. Dorje (ed.), 181–184, Boston: Wisdom Publications.

Gombrich, Richard 1975: "Ancient Indian Cosmology", in *Ancient Cosmologies*, C. Blacker and M. Loewe (ed.), 110-142, London: George Allen and Unwin.

— 1988: *Theravāda Buddhism, A Social History from Ancient Benares to Modern Colombo*, London: Routledge & Kegan Paul.

Greenberg, Jeff, T. Pyszczynski, and S. Solomon 1986: "The Causes and Consequences of a Need for Self-esteem: A Terror Management Theory", in *Public Self and Private Self*, R. F. Baumeister (ed.), 189–212, New York: Springer.

Harris, Elizabeth J. 1997a: *Detachment and Compassion in Early Buddhism*, Kandy: Buddhist Publication Society (online version).

Harris, Ian 1991: "How Environmentalist Is Buddhism?", *Religion*, 21.2: 101–114.

— 1994: "Causation and Telos: The Problem of Buddhist Environmental Ethics", *Journal of Buddhist Ethics*, 1: 45–56.

— 1995a: "Buddhist Environmental Ethics and Detraditionalization: The Case of Eco-Buddhism", *Religion*, 25: 199–211.

— 1995b: "Getting to Grips with Buddhist Environmentalism: A Provisional Typology", *Journal of Buddhist Ethics*, 2: 173–190.

— 1997b: "Buddhism and the Discourse of Environmental Concern: Some Methodological Problems Considered", in *Buddhism and Ecology, The Interconnection of Dharma and Deeds*, M. E. Tucker and D. R. Williams (ed.), 377–402, Cambridge: Harvard University Press.

Harvey, Peter 2000/2005: *An Introduction to Buddhist Ethics*, Cambridge: Cambridge University Press.

— 2007: "Avoiding Unintended Harm to the Environment and the Buddhist Ethic of Intention", *Journal of Buddhist Ethics*, 14: 1–34.

— 2013: "Buddhist Reflections on 'Consumer' and 'Consumerism'", *Journal of Buddhist Ethics*, 20: 334–356.

Hirabayashi Jiro 2015: "The Sanskrit Fragments Or. 15009/401–450 in the Hoernle Collection", in *Buddhist Manuscripts from Central Asia, The British Library Sanskrit Fragments, Volume II*, Karashima S., Nagashima J. and K. Wille (ed.), 273–313, Tokyo: International Research Institute for Advanced Buddhology, Soka University.

Hoggett, Paul 2011: "Climate Change and the Apocalyptic Imagination", *Psychoanalysis, Culture and Society*, 16.3: 261–275.

Holder, John J. 2007: "A Suffering (But Not Irreparable) Nature: Environmental Ethics from the Perspective of Early Buddhism", *Contemporary Buddhism*, 8.2: 113–130.

Horner, I.B. 1938/1982: *The Book of the Discipline (Vinaya-Piṭaka), Volume I*, London: Pali Text Society.

— 1952/1975: *The Book of the Discipline (Vinaya-Piṭaka), Volume V*, London: Pali Text Society.

Hosoda Noriaki 2003: "[A Study on the Mārgavarga of the Saṃyuktāgama: In Search of the Lost 'Prahāṇa-saṃyukta' in the Original vol. 25]", *Tohogaku*, 105: 179–165.

Ives, Christopher 2013: "Resources for Buddhist Environmental Ethics", *Journal of Buddhist Ethics*, 20: 541–571.

Jacobson Stefan Gaarsmand (ed.) 2018: *Climate Justice and the Economy, Social Mobilization, Knowledge and the Political*, New York: Routledge.

Jafry, Tahseen (ed.) 2019: *Routledge Handbook of Climate Justice*, New York: Routledge.

Jayawickrama, N.A. 1990: *The Story of Gotama Buddha, The Nidāna-kathā of the Jātakaṭṭhakathā*, Oxford: Pali Text Society.

Jing Yin 2009: "Devadatta's Personality and the Schism", in *Buddhist and Pali Studies in Honour of the Venerable Professor Kakkapalliye Anuruddha*, K.L. Dhammajoti and Y. Karunadasa (ed.), 369–392, Hong Kong: Centre of Buddhist Studies, University of Hong Kong.

Jodoin, Sébastien, R. Faucher, and K. Lofts 2019: "Look Before You Jump, Assessing the Potential Influence of the Human Rights Bandwagon on Domestic Climate Policy", in *The Routledge Handbook of Human Rights and Climate*

Governance, S. Duyck, S. Jodoin, and A. Johl (ed.), 167–182, New York: Routledge.

Karmapa, Orgyen Trinley Dorje 2009: "Pure Aspiration, Bodhisattva Activity, and a Safe-Climate Future", in *A Buddhist Response to the Climate Emergency*, J. Stanley, D. R. Loy, and G. Dorje (ed.), 81–84, Boston: Wisdom Publications.

— 2013: *The Heart is Noble, Changing the World from the Inside Out*, Boston: Shambala Publications.

Kaza, Stephanie 2008: *Mindfully Green, A Personal and Spiritual Guide to Whole Earth Thinking*, Boston: Shambala.

— 2014: "Buddhist Contributions to Climate Response", *The Journal of Oriental Studies*, 24: 73–92.

— 2018: "Buddhist Environmental Ethics: An Emergent and Contextual Approach", in *Oxford Handbook of Buddhist Ethics*, D. Cozort and J. M. Shields (ed.), 432–452, Oxford: Oxford University Press.

Keith, A.B. 1917: "Numbers (Aryan)", in *Encyclopædia of Religion and Ethics, Volume IX*, J. Hastings (ed.), 407–413, Edinburgh: T. & T. Clark.

Keown, Damien 2007: "Buddhism and Ecology: A Virtue Ethics Approach", *Contemporary Buddhism*, 8: 97–112.

Klepp, Silja and L. Chavez-Rodriguez 2018: "Governing Climate Change, The Power of Adaptation Discourses, Policies, and Practices", in *A Critical Approach to Climate Change Adaptation*, S. Klepp and L. Chavez-Rodriguez (ed.), 3–34, New York: Routledge.

Klinsky, Sonja 2019: "Beyond the Academy, Reflecting on Public Scholarship About Climate Justice", in *Routledge Handbook of Climate Justice*, T. Jafry (ed.), 467–478, New York: Routledge.

Kloetzli, W. Randolph 1983: *Buddhist Cosmology, Science and Theology in the Images of Motion and Light*, Delhi: Motilal Banarsidass.

Knoblauch, Christian, C. Beer, S. Liebner, M. N. Grigoriev, M.N. and E.-M. Pfeiffer 2018: "Methane Production as Key to the Greenhouse Gas Budget of Thawing Permafrost", *Nature Climate Change*, 8: 309–312.

Lamotte, Étienne 1970a: "Le Buddha insulta-t-il Devadatta?", *Bulletin of the School of Oriental and African Studies*, 33: 107–115.

Lamotte, Étienne 1970b: *Le Traité de la Grande Vertu de Sagesse de Nāgārjuna (Mahāprajñāpāramitāśāstra), Tome III*, Louvain-la-Neuve: Institut Orientaliste.

— 1976: *Le Traité de la Grande Vertu de Sagesse de Nāgārjuna (Mahāprajñāpāramitāśāstra), Tome IV*, Louvain-la-Neuve: Institut Orientaliste.

Levman, Bryan 2019: "The Historical Buddha: Response to Drewes", *Canadian Journal of Buddhist Studies*, 14: 25–56.

Li Channa 2019: "Devadatta", in *Brills' Encyclopedia of Buddhism, Volume II: Lives*, J. A. Silk, R. Bowring, V. Eltschinger, and M. Radich (ed.), 141–155, Leiden: Brill.

Light, Andrew 2002: "Contemporary Environmental Ethics, From Metaethics to Public Philosophy", *Metaphilosophy*, 33.4: 426–449.

Loy, David R. 2018: *Ecodharma, Buddhist Teachings for the Ecological Crisis*, Boston: Wisdom Publications.

Lueddeke, Georg R. 2019: *Survival: One Health, One Planet, One Future*, New York: Routledge.

Maithrimurthi, Mudagamuwe 1999: *Wohlwollen, Mitleid, Freude und Gleichmut, Eine ideengeschichtliche Untersuchung der vier apramāṇas in der buddhistischen Ethik und*

Spiritualität von den Anfängen bis hin zum frühen Yogācāra, Stuttgart: Franz Steiner.

Markus, Till, Bh. Vivekānanda, and M. Lawrence 2018: "An Assessment of Climate Engineering from a Buddhist Perspective", *Journal for the Study of Religion, Nature and Culture*, 12.1: 8–33.

Martini, Giuliana 2012: "The 'Discourse on Accumulated Actions' in Śamathadeva's Abhidharmakośopāyikā", *The Indian International Journal of Buddhist Studies*, 13: 49–79.

McDonald, Rachel I., H. Y. Chai, and B. R. Newell 2015: "Personal Experience and the 'Psychological Distance' of Climate Change: An Integrated Review", *Journal of Environmental Philosophy*, 44: 109–118.

Melzer, Gudrun 2006: *Ein Abschnitt aus dem Dīrghāgama*, PhD thesis, München: Ludwig-Maximilians-Universität.

Mittal, Kusum 1957: *Dogmatische Begriffsreihen im älteren Buddhismus, I, Fragmente des Daśottarasūtra aus zentralasiatischen Sanskrit-Handschriften*, Berlin: Akademie Verlag.

Moo, Jonathan 2015: "Climate Change and the Apocalyptic Imagination: Science, Faith, and Ecological Responsibility", *Zygon*, 50.4: 937–948.

Mukherjee, Biswadeb 1966: *Die Überlieferung von Devadatta dem Widersacher des Buddha in den kanonischen Schriften*, München: Kitzinger.

Nagashima Jundo 2015: "The Sanskrit Fragments Or. 15009/ 501–600 in the Hoernle Collection", in *Buddhist Manuscripts from Central Asia, The British Library Sanskrit Fragments, Volume II*, Karashima S., Nagashima J. and K. Wille (ed.), 347–418, Tokyo: International Research Institute for Advanced Buddhology, Soka University.

Nakamura Hajime 2000: *Gotama Buddha, A Biography Based on the Most Reliable Texts*, vol. 2, Tokyo: Kosei Publishing Co.

Ñāṇamoli, Bhikkhu 1991: *The Path of Purification (Visuddhimagga) by Bhadantācariya Buddhaghosa*, Kandy: Buddhist Publication Society.

— 1995/2005: *The Middle Length Discourses of the Buddha, A Translation of the Majjhima Nikāya*, Bhikkhu Bodhi (ed.), Boston: Wisdom Publications.

Ñāṇaponika, Thera 1966/1981: *The Greater Discourse on the Elephant-Footprint Simile, From the Majjhima Nikāya*, Kandy: Buddhist Publication Society.

— 1990: *Protection Through satipaṭṭhāna*, Kandy: Buddhist Publication Society.

Nattier, Jan 1991: *Once Upon a Future Time: Studies in a Buddhist Prophecy of Decline*, Berkeley: Asian Humanities Press.

Nhat Hanh, Thich 2008: *The World we Have, A Buddhist Approach to Peace and Ecology*, Berkeley: Parallax Press.

Norman, K.R. 1969: *The Elder's Verses I, Theragāthā, Translated with an Introduction and Notes*, London: Pali Text Society.

— 1990/1993: "Pāli Philology and the Study of Buddhism", in *Collected Papers Volume IV*, K.R. Norman (ed.), 81–91, Oxford: Pali Text Society.

— 1997/2004: *The Word of the Doctrine (Dhammapada)*, Oxford: Pali Text Society.

Ohnuma Reiko 2017: *Unfortunate Destiny, Animals in the Indian Buddhist Imagination*, New York: Oxford University Press.

Otto, Daniel, S. Caeiro, P. Nicolau, A. Disterheft, A. Teixeira, S. Becker, A. Bollmann, and K. Sander 2019: "Can MOOCs Empower People to Critically Think About Climate Change?

A Learning Outcome Based Comparison of Two MOOCS",
Journal of Cleaner Production, 222: 12–21.

Pradhan, P. 1967: *Abhidharmakośabhāṣya of Vasubandhu*, Patna: K.P. Jayaswal Research Institute.

Pruden, Leo M. 1988: *Abhidharmakośabhāṣyam by Louis de la Vallée Poussin*, Berkeley: Asian Humanity Press.

Ranney, Michael Andrew and D. Clark 2016: "Climate Change Conceptual Change: Scientific Information Can Transform Attitudes", *Topics in Cognitive Science*, 8: 49–75.

Ray, Reginald A. 1994: *Buddhist Saints in India, A Study in Buddhist Values & Orientations*, New York: Oxford University Press.

Rhys Davids, T.W. and C.A.F. Rhys Davids 1921 (vol. 3): *Dialogues of the Buddha, Translated from the Pali of the Dîgha Nikâya*, London: Oxford University Press.

Rhys Davids, T.W. and W. Stede 1921/1993: *Pali-English Dictionary*, Delhi: Motilal Banarsidass.

Ringu Tulku Rinpoche 2009: "The Bodhisattva Path at a Time of Crisis", in *A Buddhist Response to the Climate Emergency*, J. Stanley, D. R. Loy, and G. Dorje (ed.), 127–134, Boston: Wisdom Publications.

Rockefeller, Steven C. 1997: "Buddhism, Global Ethics, and the Earth Charter", in *Buddhism and Ecology, The Interconnection of Dharma and Deeds*, M. E. Tucker and D. R. Williams (ed.), 313–324, Cambridge: Harvard University Press.

Sadakata Akira 1997/2004: *Buddhist Cosmology, Philosophy and Origins*, G. Sekimori (trsl.), Tokyo: Kōsei Publishing.

Sandell, Klas 1987: "Buddhist Philosophy as Inspiration to Ecodevelopment", in *Buddhist Perspectives on the Ecocrisis*, K. Sandell (ed.), 42–51, Kandy: Buddhist Publication Society (online version).

Sander, Lore and E. Waldschmidt 1985: *Sanskrithandschriften aus den Turfanfunden, Teil 5*, Wiesbaden: Franz Steiner.

Schmithausen, Lambert 1991: *The Problem of the Sentience of Plants in Earliest Buddhism*, Tokyo: International Institute for Buddhist Studies.

— 1997a: "The Early Buddhist Tradition and Ecological Ethics", *Journal of Buddhist Ethics*, 4: 1–74.

— 1997b: *Maitrī and Magic: Aspects of the Buddhist Attitude Towards the Dangerous in Nature*, Wien: Verlag der Österreichischen Akademie der Wissenschaft.

— 2000: "Buddhism and the Ethics of Nature, Some Remarks", *The Eastern Buddhist*, 32.2: 26–78.

— 2009: *Plants in Early Buddhism and the Far Eastern Idea of the Buddha-Nature of Grasses and Trees*, Lumbini: Lumbini International Research Institute.

Schröder, K.-P. and R. C. Smith 2008: "Distant Future of the Sun and Earth Revisited", *Monthly Notices of the Royal Astronomical Society*, 386.1: 155–163.

Seeger, Martin 2014: "Ideas and Images of Nature in Thai Buddhism: Continuity and Change", in *Environmental and Climate Change in South and Southeast Asia, How are Local Cultures Coping?*, B. Schuler (ed.), 43–74, Leiden: Brill.

Sen, Benoychandra 1974: *Studies in the Buddhist Jātakas (Tradition and Polity)*, Calcutta: Saraswat Library.

Senart, Émile 1882: *Essai sur la légende du Buddha, son caractère et ses origines*, Paris: Ernest Leroux.

Shakhova, Natalia, I. Semiletov, and E. Chuvilin 2019: "Understanding the Permafrost-Hydrate System and Associated Methane Releases in the East Siberian Arctic Shelf", *Geosciences*, 9.251; 1–23.

Silverlock, Blair Alan 2015: *An Edition and Study of the Gosiga-sutra, the Cow-horn Discourse (Senior Collection*

Scroll no. 12): An Account of the Harmonious Aṇarudha Monks, PhD dissertation, University of Sydney.

Sivaraksa, Sulak 2009: *The Wisdom of Sustainability, Buddhist Economics for the 21st Century*. Kihei: Koa Books.

Smith, Bruse W., C. Graham Ford, and Laurie E. Steffen 2019: "The Role of Mindfulness in Reactivity to Daily Stress in Urban Firefighters", *Mindfulness*, 10: 1603–1614.

Sponberg, Alan 1997: "Green Buddhism and the Hierarchy of Compassion", in *Buddhism and Ecology, The Interconnection of Dharma and Deeds*, M. E. Tucker and D. R. Williams (ed.), 351–376, Cambridge: Harvard University Press.

Stanescu, Vasile 2020: "'Cowgate', Meat Eating and Climate Change Denial", in *Climate Change Denial and Public Relations, Strategic Communication and Interest Groups in Climate Action*, N. Almiron and J. Xifra (ed.), 178–194, New York: Routledge.

Strain, Charles 2016: "Engaged Buddhist Practice and Ecological Ethics: Challenges and Reformulations", *Worldviews–Global Religions Culture and Ecology*, 20.2: 189–210.

Stuart, Daniel 2019: "Becoming Animal: Karma and the Animal Realm Envisioned through an Early Yogācāra Lens", *Religions*, 10.363: 1–15.

Sucitto, Ajahn 2019: *Buddha-Nature, Human Nature*, Great Gaddesden: Amaravati Publications.

Swearer, Donald K. 2001: "Principles and Poetry, Places and Stories: The Resources of Buddhist Ecology", *Daedalus*, 130.4: 225–241.

— 2006: "An Assessment of Buddhist Eco-Philosophy", *Harvard Theological Review*, 99.2: 123–137.

Syrkin, A. 1983: "On the First Work in the Sutta Piṭaka: The Brahmajāla-Sutta", in *Buddhist Studies: Ancient and Modern*, P. Denwood (ed.), 153–166, London: Curzon.

Thanissara 2015: *Time to Stand Up, An Engaged Buddhist Manifesto for Our Earth, The Buddha's Life and Message through Feminine Eyes*, Berkeley: North Atlantic Books.

Tokar, Brian 2018: "On Social Ecology and the Movement for Climate Justice", in *Climate Justice and the Economy, Social Mobilization, Knowledge and the Political*, S. G. Jacobson (ed.), 168–187, New York: Routledge.

Tripāṭhī, Chandrabhāl 1962: *Fünfundzwanzig sūtras des Nidānasaṃyukta*, Berlin: Akademie Verlag.

Troillet, Mélodie, T. Barbier, and J. Jacquet 2019: "From Awareness to Action: Taking Into Consideration the Role of Emotions and Cognition for a Stage Toward a Better Communication of Climate Change", in *Addressing the Challenges in Communicating Climate Change Across Various Audiences*, W. L. Filho, B. Lackner, and H. McGhie (ed.), 47–64, Cham: Springer.

Vansina, Jan 1985: *Oral Tradition as History*, Madison: University of Wisconsin Press.

Visigalli, Paolo 2019: "Chartering 'Wilderness' (araṇya) in Brahmanical and Buddhist Texts", *Indo-Iranian Journal*, 62: 162–189.

Waldschmidt, Ernst 1944: *Die Überlieferung vom Lebensende des Buddha, Eine vergleichende Analyse des Mahāparinirvāṇasūtra und seiner Textentsprechungen, Erster Teil*, Göttingen: Vandenhoeck & Ruprecht.

Wallace-Wells, David 2019: *The Uninhabitable Earth, Life After Warming*, New York: Tim Duggan Books.

Walshe, Maurice 1987: *Thus Have I Heard; The Long Discourses of the Buddha*, London: Wisdom Publications.

Wille, Klaus 2008: *Sanskrithandschriften aus den Turfanfunden, Teil 10*, Stuttgart: Franz Steiner.

Willemen, Charles 1978: *The Chinese Udānavarga, A Collection of Important Odes of the Law, Fa Chi Yao Sung Ching,*

Translated and Annotated, Bruxelles: Institut Belge des Hautes Études Chinoises.

— 1999: *The Scriptural Text: Verses of the Doctrine, With Parables, Translated from the Chinese of Fa-li and Fa-chü*, Berkeley: Numata Center for Buddhist Translation and Research.

Wilson, Jeff 2014: *Mindful America, The Mutual Transformation of Buddhist Meditation and American Culture*, Oxford: Oxford University Press.

Wright, J.C. 2000: "Pāli dīpam attano and attadīpa", in *Harānandalaharī, Volume in Honour of Professor Minoru Hara on His Seventieth Birthday*, R. Tsuchida and A. Wezler (ed.), 481–503, Reinbek: Dr. Inge Wezler Verlag für Orientalistische Fachpublikationen.

Wujastyk, Dominik 2004: "Jambudvīpa: Apples or Plumes?", in *Studies in the History of the Exact Sciences in Honour of David Pingree*, C. Burnett, J. P. Hogendijk, K. Plofker, and M. Yano (ed.), 287–301, Leiden: Brill.

Yuen, Eddie 2012: "The Politics of Failure Have Failed: The Environmental Movement and Catastrophism", in *Catastrophism: The Apocalyptic Politics of Collapse and Rebirth*, L. Sasha, D. McNally, E. Yuen, and J. Davis (ed.), 15-43, Oakland: PM Press.

Zhang Tieshan 2002: "An Uighur Fragment of the Za-ahanjing from Dunhuang Preserved in the Library of the Beijing University", *Journal of the Central University for Nationalities, Philosophy and Social Sciences Edition*, 4: 108–112.

Index

Also by Bhikkhu Anālayo:

Bhikkhunī Ordination from Ancient India to Contemporary Sri Lanka
Buddhapada and the Bodhisattva Path
A Comparative Study of the Majjhima-nikāya
Compassion and Emptiness in Early Buddhist Meditation
The Dawn of Abhidharma
Dīrgha-āgama Studies
Early Buddhist Meditation Studies
Ekottarika-āgama Studies
Excursions into the Thought-world of the Pāli Discourses
The Foundation History of the Nuns' Order
The Genesis of the Bodhisattva Ideal
Madhyama-āgama Studies
A Meditator's Life of the Buddha
Mindfully Facing Disease and Death
Mindfulness of Breathing
Perspectives on Satipaṭṭhāna
Rebirth in Early Buddhism and Current Research
Saṃyukta-āgama Studies
Satipaṭṭhāna, the Direct Path to Realization
Satipaṭṭhāna Meditation: A Practice Guide
Vinaya Studies

About the
Barre Center for Buddhist Studies

The Barre Center for Buddhist Studies is a non-profit educational organization dedicated to exploring Buddhist thought and practice as a living tradition, faithful to its origins, yet adaptable to the current world. The Center, located in Barre, Massachusetts, offers residential and online courses combining study, discussion, and meditation for the purpose of deepening personal practice while building and supporting communities of like-minded practitioners. Our programming is rooted in the classical Buddhist tradition of the earliest teachings and practices, but calls for dialogue with other schools of Buddhism and with other academic fields—and with each other. All courses support both silent meditation practice and critical, dialogical investigation of the teachings.

BCBS was founded in 1991 by teachers at Insight Meditation Society (IMS), including Joseph Goldstein and Sharon Salzberg, and is connected to both IMS and the Forest Refuge by trails through the woods.

BCBS is committed to cultivating a community that reflects the diversity of our society and our world. We seek to promote the inclusion, equity and participation of people of diverse identities so that all may feel welcome, safe, and respected within this community. Find out more about our mission, our programs, and sign up for our free monthly e-newsletter at www.buddhistinquiry.org.

BARRE CENTER *for*
BUDDHIST STUDIES